Emotional Longevity

What *Really* Determines How Long You Live

Norman B. Anderson, Ph.D.,

with P. Elizabeth Anderson

VIKING

VIKING
Published by the Penguin Group
Penguin Putnam Inc., 375 Hudson Street,
New York, New York 10014, U.S.A.
Penguin Books Ltd, 80 Strand,
London WC2R 0RL, England
Penguin Books Australia Ltd, 250 Camberwell Road, Camberwell,
Victoria 3124, Australia
Penguin Books Canada Ltd, 10 Alcorn Avenue,
Toronto, Ontario, Canada M4V 3B2
Penguin Books India (P) Ltd, 11 Community Centre, Panchsheel Park,
New Delhi – 110 017, India
Penguin Books (N.Z.) Ltd, Cnr Rosedale and Airborne Roads, Albany,
Auckland, New Zealand
Penguin Books (South Africa) (Pty) Ltd, 24 Sturdee Avenue,
Rosebank, Johannesburg 2196, South Africa

Penguin Books Ltd, Registered Offices:
Harmondsworth, Middlesex, England

First published in 2003 by Viking Penguin,
a member of Penguin Putnam Inc.

1 3 5 7 9 10 8 6 4 2

Every effort has been made to ensure that the information contained in this book is complete and accurate. However, neither the publisher nor the author is engaged in rendering professional advice or services to the individual reader. The ideas, procedures, and suggestions contained in this book are not intended as a substitute for consulting with your physician. All matters regarding your health require medical supervision. Neither the author nor the publisher shall be liable or responsible for any loss, injury, or damage allegedly arising from any information or suggestion in this book.

LIBRARY OF CONGRESS CATALOGING-IN-PUBLICATION DATA
Anderson, Norman B.
Emotional longevity : what *really* determines how long you live / Norman B. Anderson
with P. Elizabeth Anderson.
p. cm.
Includes bibliographical references and index.
ISBN 0-670-03185-2
1. Longevity. 2. Emotions. 3. Aging—Prevention. I. Anderson, P. Elizabeth. II. Title.
RA776.75 .A534 2003
612.6'8—dc21   2002033071

This book is printed on acid-free paper. ∞

Printed in the United States of America
Set in Meridien
Designed by BTDNYC

This book is dedicated to my mother,
the late Reverend Dr. Lois Anna Jones Anderson,
who was the personification of health
in all of its dimensions.

Preface and
Acknowledgments

The motivation to write this book grew out of one of the most extraordinary professional experiences of my life. In 1995, I was hired as the first associate director of the National Institutes of Health (NIH) and the founding director of its Office of Behavioral and Social Sciences Research, or OBSSR. I was honored to have the chance to work at NIH, since it is truly one of our national treasures. For those less familiar with it, the NIH is the federal government's main health-research arm; it is composed of twenty-seven institutes and centers, with a budget in 2002 of approximately $23 billion. That budget is distributed across the twenty-seven institutes and centers at NIH that fund and conduct research in areas such as cancer, mental health, heart disease, aging, diabetes, and musculoskeletal disorders, among others. The largest percentage of budgets from these institutes and centers goes to university researchers in the form of grants to conduct health-related research. Most of the health breakthroughs you read about in the newspaper or hear about on the evening news were likely supported by grants from the NIH.

The late 1990s were an exciting time to work at NIH. The institution was headed by Nobel Laureate Dr. Harold Varmus, who was appointed by President Clinton in 1993 and who immediately raised the intellectual bar at a place that was already the best in the world at what it did. The Human Genome Project was in full swing and nearing completion, and discoveries about the genetics of disease and the potential for new medical treatments were emerging at a dizzying pace. Congress was impressed with NIH and its work, making the institution one of the few federal agencies for which there was wide consensus that its budget actually needed to *grow,* rather than shrink. In a remarkable show of bipartisan unanimity, Congress and the Clinton administration set into motion a plan to double the NIH budget between 1999 and 2003, a plan whose implementation has continued under President Bush.

By its nature, health science is a multidisciplinary endeavor, meaning that no one discipline "owns" the field. Obviously medicine has a large stake in it, and of course the basic biological sciences such as molecular biology (genetics), biochemistry, and physiology are central figures. These fields and others, which can collectively be called biomedicine, are the dominant areas of NIH and health research. But over the last few decades there have been major contributions to understanding the nature of health and illness from scientists from the behavioral and social sciences, including disciplines such as psychology, sociology, public health, demography, anthropology, and others. As director of OBSSR, my job, as mandated by Congress, was to work within NIH to boost research efforts concerning these latter fields and to integrate the work being conducted in biomedicine with that stemming from the behavioral and social sciences. A large part of my job was communicating to—in fact, trying to convince—many biomedical scientists, administrators, policy makers, and laypeople that behavioral and social aspects of health were worthy of federal investments. It is not that the health relevance of behaviors such as smoking, physical activity, diet, and taking medications properly was not recognized, but that there was a perception that behavioral and social sciences did not have more to offer health science. The era of molecular biology was progressing so rapidly that I even heard someone remark that someday there would be no need for research on behavioral aspects of health. Complicating my job further was the fact that many biomedical, behavioral, and social scientists perceive their work as the blind men in John Godfrey Saxe's famous poem perceived an elephant— each man's image of the elephant reflected which part of it he was touching. The man touching the tail described the elephant much differently than the ones holding the trunk or knee. Scientists working in different disciplines often "see" health problems from the perspective of that discipline, frequently unaware of relevant research on the same problems coming from other fields. A geneticist might say that to really understand heart disease, you have to study genes, while a psychologist might say that to really understand heart disease, you have to study stress. But both processes affect heart disease and, in fact, stress may trigger the expression of certain genes. So, the two processes are inextricably linked, although we study them as if they were independent of each other.

One of my main tasks at NIH was to foster connections and collaborations between biomedical and behavioral and social scientists while making the strongest case for the relevance of the latter to health science. This book is really about the research that has actually done both of these things and that has led to what I call a new health science. The new health science has literally provided the empirical justification for a truly comprehensive view of the many dimensions of health, which are depicted on the jacket of this book and in the introduction. This new health science brings all of the components of the health "elephant" together, so that for the first time we can see, as the subtitle of this book indicates, what *really* determines how long you live and what really underlies the chronic illnesses that plague our society.

Although this is a book about this exciting new health science, it is written explicitly for the nonscientist. I certainly hope scientists will read it too, but I wanted to bring to a lay audience some of the intriguing discoveries that have resulted from the new health science, but do so in a nontechnical way that nevertheless remained true to the complexities of the research. Research in the new health science is vast, which meant that I had to be selective in choices I made about what to include. I apologize to my colleagues whose areas of research were not included and to those whose specific research was not mentioned. What kept me awake at night the most while writing the book was the knowledge of all the great research and discoveries that were not included, many of which were just published when this book was going to press. My collaborator and spouse, P. Elizabeth, and I have developed a Web site that will continue to provide information about more recent findings (www.emotionallongevity.com). I also strongly recommend the Web site for the Center for the Advancement of Health (www.cfah.org), which is dedicated to advancing the very ideas presented here. The center's Web site is among the best on the topic.

There are many people I need to thank for making this book possible. The book would not exist without the expertise, guidance, advice, and friendship from two key people. Our literary agent, Ms. Jenny Bent, who worked tirelessly with Elizabeth and me to shape an acceptable book proposal, and our editor at Viking, Ms. Molly Stern, who was willing to take on two novice book authors and whose patience and perseverance, editorial talents, and motivational skills made this final product possible. I am also grateful to Dr. Barbara Lynch, who read and rigorously critiqued an early version of this manuscript.

I wrote this book after I left NIH and while I was a professor at the Harvard University School of Public Health. I want to express my deep appreciation to Dr. Lisa Berkman, chair of the Department of Health and Social Behavior where I taught, and Deans Barry Bloom and James Ware, all of whom provided me with the time and support I needed to complete this project. Special thanks also goes to my assistant, Maribel Herrera, and to the faculty, students, and staff of the Department of Health and Social Behavior at Harvard, whose collegiality and encouragement were greatly valued.

My thanks to the many mentors and colleagues I have had over the years—through graduate school, clinical internships, postdoctoral fellowships, and faculty appointments—who introduced me to and excited my interest in research at the nexus of behavioral and biomedical science. I am especially grateful to my scientific "father," Dr. Redford Williams, and to my other mentors and colleagues at Duke University School of Medicine, especially Drs. Francis Keefe, Richard Surwit, James Blumenthal, and James Lane. Special thanks also to Drs. Michael Follick, David Abrams, and David Ahern at Brown University School of Medicine, and to Dr. Neil Schneiderman of the University of Miami and Drs. Sherman James and James Jackson of the University of Michigan, who have been a constant source of encouragement and friendship since the start of my career.

Special thanks to Dr. Harold Varmus, former head of NIH and now president of Memorial Sloan-Kettering Cancer Center, for giving me the opportunity to work at NIH and serve during his administration there, and to former NIH Deputy Director Dr. Ruth Kirschstein for her guidance and support throughout my NIH tenure. I am particularly grateful to my former staff at OBSSR and to the more than five hundred behavioral and social science administrators and researchers who work at NIH who embraced me from day one and helped make OBSSR a success. These individuals form an all too infrequently recognized foundation for all of the research described in this book.

A book about science would not be possible without excellent science to write about, so I am most grateful to the hundreds of scientists who work in this field and whose research is changing our views on what it means to be healthy. I am a little biased, but I believe the level of rigor and creativity that scientists in the new health science bring to the research enterprise is unparalleled. Often these researchers are addressing concepts that can be difficult to measure, like emotions, stress, and social relationships, and where they must demonstrate the connection between such concepts to biology, illnesses, and death in ways that pass the highest tests of scientific rigor. As I hope you will learn by reading this book, they have done an amazing job of doing just that.

I want to especially thank those scientists whose research is featured in the book and who took the time to answer my e-mail and telephone queries about their work and who read sections of the book. Although I cover a lot of scientific territory and am not myself an expert in all the fields I discuss, I tried my best to present the work accurately and to "get it right." I apologize to those who might feel I fell short of this goal in describing their research.

Although this is a book about science, it is the stories of the lives of real people that bring the science to life. I am grateful for those who shared their stories with Elizabeth and me through interviews and for those whose stories where gleaned from other sources.

My highest gratitude goes to my collaborator and spouse, Elizabeth, without whose expertise, emotional support, and love this project would have collapsed long ago. She is the only one who helped me shoulder the weight of this project from day one and who had a front-row seat for the good, the bad, and the ugly of my moods about it. When one of us felt like giving up, the other one was there to say the right thing to provide a ray of hope that this book would somehow get done (and there were many times we frankly weren't sure). It is often said that you should never work on a major project with your spouse. But working on this book together, while not always easy, allowed us to deepen and broaden our relationship and our love for each other in ways that otherwise would not have been possible. And for that I am grateful.

Norman B. Anderson, Ph.D.

Norman obviously did the heavy lifting for *Emotional Longevity*, but sometimes my part seemed overwhelming. I must thank the people who helped me hold up my end. First, my mother, Marguerite Leggett, who in her eighties is another picture of longevity. Then the friends whom I neglected for the nearly two years. Barbara Anne Harvey, my sister, for being a reading machine. Barbara Cooke, who understands everything. Ann T. Fico, LCMT, for giving me a sanctuary and not letting my body give out. Peter Phipps, my editor at the *Providence Journal*, for being a kind, compassionate, trustworthy soul—a mensch and the best boss ever. Jenny Bent, our agent—the first other person to believe in *Emotional Longevity*—a gift from heaven who became a cherished friend. Molly Stern, our editor at Penguin Putnam, for her boundless enthusiasm, for her steadfast belief in our dream, and for her priceless inspiration. This would not be the book it is without her. Dr. Barbara Lynch, a friend and colleague, for her invaluable insights and comments. Dr. Maya Angelou, for wise words at a crucial time. Everyone who prayed for us. Last, but never least, my husband, Norman, whose soul recognized mine sixteen years ago, giving me a wonderful life.

P. Elizabeth Anderson

Contents

List of illustrations

Introduction

Emotional Longevity: Toward a New Definition of Health

The greatest compliment I have ever received came from my mother, in her last words, on the last day of her life. She said, "You are becoming just like me."

As a son I've often pondered the meaning and significance of what she said, as I've pondered her long and vital life. What did it mean to be just like her? I will perhaps never fully understand the meaning *she* intended by her statement, but I do know that my mother possessed many characteristics worthy of emulation. Among the most impressive was the harmony of her moods, where anger or fear were rarely evident and sadness was short-lived, but where contentment, joy, and happiness were nearly ever present, even when external circumstances were less than uplifting.

As a psychologist and a health researcher, I have also pondered how science could uncover what contributed to my mother's emotional symmetry, and her longevity, and make it available for everyone. To uncover in a sense what it means to be truly whole, in body, mind, and spirit, and package it for all. Although this is a tall order for science, I think we are actually getting close—closer than we've ever been before. This book is about a new revolution in health science that is getting us there, and in doing so is literally transforming what it means to be healthy and demonstrating how we can enhance the quality and length of our lives.

These days when you hear about a revolution in health science, you are in all probability hearing about the *genetics* revolution, and for good reason. Genetics has already transformed biological science and has the promise to change the practice of medicine. But health science is also experiencing another revolution—one that is less publicized but is, in some respects, as complex and profound as the recent breakthroughs in genetics. This other revolution, like the one in genetics, is fueled by rigorous scientific methodology and advanced statistical analyses. But, unlike the genetics revolution, this one can be brilliantly illuminated in the lives of everyday people. People who have the kinds of lives most of us strive for—lives that are happy, fulfilled, confident, mentally engaged, and vital. These are people who are able to hold off sickness until the very end of long lives and who do not suffer through years of chronic illness. If illness comes, these are people who recover faster or who, like my mother, are able to maintain a sense of emotional and psychological well-being, and perhaps even grow from the experience.

My mother, Lois Anna Jones Anderson, died in 1992 at the age of seventy-eight. Although she was ultimately claimed by cancer, the arc of her life and her senior years were characterized by strong health and vigor. For nearly fifty years she and my father served as copastors of a sizable Baptist church in North Carolina, making her one of the first women leaders of a large Southern congregation. Though leading a congregation was not always heaven on earth, my mother reveled in the rich and rewarding life of the church. It provided her an intellectual outlet, where she prepared weekly sermons and Bible-study lessons. Her personal relationships were broad and deep, not only with members of the church but also with many people in the community. These people provided her with a vast network of mutual support. My mother also had a very keen mind and loved to read on a variety of topics, but she especially loved books on spiritual matters. When she married my father, she had only finished high school, but she valued education so much that she squeezed in college courses between church duties and raising a family. She earned a bachelor's and a master's degree and was ultimately awarded an honorary doctorate in theology.

My mother's life was not perfect, and she had her share of negative experiences. At the age of eight, she was devastated by the death of her own mother. Two years later her younger sister died of an infection after accidentally sticking a pencil in her eye. While still coping with those deaths, she lost the security of her home in Norfolk, Virginia, when her father abruptly moved the family to New York City. She suffered several miscarriages trying to have children. Much later in life my mother coped with several debilitating chronic illnesses in the family, from Alzheimer's disease in two sisters and a brother to my father's coma following a heart attack, which left him in a deep vegetative state for eighteen months. When my father died, my mother took over the reins of the two-thousand-member church alone, at the age of seventy. This was to be another trying time for her. As many historians have documented, women leaders in the church are often met with strong opposition. My mother had been a full partner with my father, providing spiritual guidance and counsel to all church members. Yet, without my father's presence, a few vocal male leaders of the congregation could not adjust to a woman at the helm. These men spearheaded a very bitter and ugly revolt, which severely divided the church. To end the turmoil, my mother, after much thought and prayer, relinquished her leadership and graciously resigned. In essence, saving the church meant losing the defining element of her life: that of a church leader at the institution she helped build, nurture, and strengthen.

It was during times of trauma and challenge that I was most impressed with my mother. She always faced obstacles with a combination of dignity, faith, wisdom, optimism, and quiet confidence. She seemed to rise above circumstances that would leave most of us reeling. In fact, it was during her ultimate battle with cancer that her remarkable attributes were most clearly manifest.

By any measure my mother had achieved what has come to be known as "successful aging," defined as experiencing minimal disease or disability until the end of life, a high level of mental and physical functioning, and an engagement in life.[1] She enjoyed a high quality of life—even in the presence of untoward circumstances—something we would all like to achieve. But is her experience realistic for most of us? Is it really possible to avoid the sickness and disability thought to inevitably accompany old age? Can we hope to achieve or maintain physical, mental, and emotional health as we grow older? Is it possible to have a positive outlook on life and experience happiness in the presence of negative life circumstances?

Science has progressed to a point where I believe that the answer to these questions is unequivocally "Yes!" New research is uncovering not only the dimensions of successful aging but also the elements of a larger phenomenon that I call *emotional longevity*. Although the expression may appear at first to connote long-lasting emotions, my meaning is very different. "Emotional longevity" is a symbolic rather than literal expression, meant to signal a departure from traditional ways of thinking about longevity. In the past we have thought of longevity, either implicitly or explicitly, as determined primarily by biology, but the phrase "emotional longevity" symbolizes a shift in emphasis away from an exclusive focus on biology. It serves to highlight the essence of the new way of thinking about longevity and health, an essence characterized by *connections*—connections between biology and social relationships; between biology and beliefs and behavior; and between biology and emotions. Science is now documenting these and other connections and showing us that attaining physical and mental health and longevity involves much more than being biologically sound, and even more than staying physically fit and eating a proper diet. Emotional longevity is the idea that these connections, between biological and nonbiological factors,[2] occurring across the life span, are what ultimately determine our health, the quality of our lives, and how long we live.

This book is about the science that has uncovered these connections and how they affect our health. It is also about people, like my mother, whose lives illustrate the new scientific findings, which together are changing our views on what it means to be healthy.

SO WHAT DOES IT MEAN TO BE HEALTHY?

The traditional answer to this question is fairly straightforward: Being healthy means an absence of disease. With no diagnosis or symptoms, traditional medical science gives us a clean bill of health. This does not mean that well-known risk factors for disease are ignored; cigarette smoking, high cholesterol, and obesity do raise concern. But in the absence of a diagnosable illness, we are still considered "healthy." Yet the absence of disease, even the absence of traditional risk factors, can be misleading.

Are We Missing Something?

In this book I discuss research that tracks large groups of individuals, often for years, who are free of disease. Invariably a certain percentage of these individuals die *prematurely* from illnesses that were not predicted based on their age, initial health, or established risk factors. If we define health simply as the absence of disease or traditional biological or behavioral risk factors, we are obviously missing something important. But what? One answer could legitimately come from genetics, since we know that certain illnesses have a strong genetic component. Yet future genetic tests will likely leave a large gap in the ability to accurately predict longevity and will further serve to highlight the need to encompass nonbiological aspects of disease.

For example, identical twins, because they share the same genetic structure, are at the same *genetic risk* for illness. Yet even twins vary tremendously in their experience of health problems such as cancer, heart disease, Alzheimer's disease, schizophrenia, Parkinson's disease, hypertension, and rheumatoid arthritis.[3] Here is a case in point: A study of the prevalence of cancer in twins, involving over 44,000 pairs of twins, was recently published in the *New England Journal of Medicine*.[4] The study was designed to estimate the relative importance of heritable and environmental factors in causing cancer. While genetic factors were clearly associated with different types of cancer in this study, they had less impact than environmental factors. The researchers noted that "inherited genetic factors make a minor contribution to susceptibility to most types" of cancer and "that the overwhelming contributor to the causation of cancer . . . was the environment."

Something other than genes alone or, more likely, something working in conjunction with genes, is determining whether one twin gets sick while his/her genetically identical sibling stays well. Animal studies have demonstrated that social, behavioral, and environmental factors can actually determine whether genes are expressed—that is, whether they are turned on or off. For example, stress has been shown to cause symptoms of diabetes, such as hyperglycemia, in animals that are genetically susceptible to diabetes. Animals not exposed to stress conditions were less likely to develop hyperglycemia or diabetes, even though they were genetically prone to the disorder.[5]

One thing is certain: Individuals determined to be at genetic risk for illness will be very motivated to do whatever is in their power to remain healthy. In the absence of gene therapies or gene-specific drugs, we will have to look for other avenues, including nonbiological ones, to keep people healthy. Health researchers are on a search for the missing components, the other nongenetic factors, that determine our health. This is a search that goes beyond even the well-known "lifestyle" factors such as diet, exercise, and smoking.

The fact that we are missing important dimensions of what constitutes health is also starkly illustrated in studies on something called *self-rated health*. Participants in these studies, who are of the same age, are asked to rate their health on a scale of one to five, where one is "excellent" and five is "poor." Results show that people who are free of disease, symptoms, and risk factors, but who nevertheless rate their own health as "fair or poor," are likely to die at an earlier age than individuals who rate their health as "good or excellent."[6] What does this mean? Why would a self-rating of poor health predict premature death in the absence of disease, symptoms, or risk factors? Were these participants tapping in to other dimensions of their own health? What was it that they sensed was awry? We can assume that they were not "sensing" a poor genetic makeup. More plausibly they were tapping in to other dimensions of health, dimensions that over time may alter their biology, leading to premature death. This book describes many remarkable studies that have not only identified these dimensions but have also conclusively linked them with biological functioning in the short term (e.g., immune and cardiovascular functioning) and with illness and longevity over the long term.

The experience of people with specific illnesses such as heart disease, cancer, or AIDS also highlights the need to broaden our definition of the determinants of health and longevity. People with such illnesses experience a wide range of outcomes. Some recover completely, with no sign of observable illness. Others, even though they continue to have the illness, are able to live productive or fulfilling lives. Yet for others the disease continues its negative progression, often leading to severe disability and death. Still others become emotionally debilitated by their illnesses. You might even say that some people with an illness are "healthier" than others with the same diagnosis. Heart disease is a perfect example. Among people who suffer heart attacks of similar severity, some recover with minimal physical impairment. Others become significantly disabled, unable to resume many previous activities. Many others go on to have subsequent heart attacks or die from the disorder sooner than their counterparts. These differences among heart patients often arise even when there are no differences in age, gender, race or ethnicity, risk factors, cardiovascular pathology, or their level of participation in rehabilitation programs. What these studies suggest, then, is that among the medically ill there are important determinants of longevity that go beyond medical features of a disease. A patient's prognosis may depend on more than the usual laundry list of biological or behavioral characteristics.

A New Definition of Health

What, then, determines the wide differences in longevity among the physically healthy and those with medical illnesses? What are the missing pieces? Over the last decades science has been inching closer to identifying these pieces. In fact, health research has come to a watershed that indicates a need to transform our previous notions and that is, in essence, leading to a new definition of health.

Specifically, the new definition of health holds that health is multifaceted and includes six fundamental dimensions, the missing pieces. As depicted in Figure 1, these dimensions include:

- Biological well-being—*biology* in the figure
- Psychological and behavioral well-being—*thoughts & actions* in the figure
- Environmental and social well-being—*environment & relationships* in the figure
- Economic well-being—*personal achievement & equality* in the figure
- Existential/religious/spiritual well-being—*faith & meaning* in the figure
- Emotional well-being—*emotions* in the figure

Your biological status is very important, but your health is more than that alone. It has a psychological and behavioral dimension that

Figure 1. THE DIMENSIONS OF OUR HEALTH

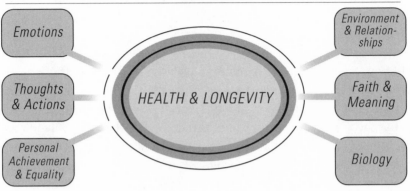

takes into account the well-known lifestyle factors ("actions") such as physical activity, diet, and smoking but also includes less recognized factors such as our expectations, explanations, and beliefs ("thoughts"); and how we respond to traumatic experiences. There is an environmental and social dimension that includes characteristics of the physical environments in which we live (e.g., level of pollution, type of housing, neighborhood safety), and our relationships with others, on both an interpersonal and societal level. The economic dimension includes not only our educational attainment and income level, but also the notion of economic equality—degree of economic difference between the most and the least well off in society. The existential/religious/spiritual dimension relates to, among other things, our beliefs and actions that help us foster a sense of meaning in our lives. For many people, meaning is fostered through religious or spiritual activities, but may also be developed through a variety of other pursuits. The final dimension, emotional well-being, involves the degree to which we experience both negative and positive emotional states. As you will read later, I believe this emotional dimension is an essential part of the "connective tissue" that links the other elements to physical well-being and longevity.

Do we really gain anything by adding these additional dimensions under the rubric of health? I argue that we do, since our lives—how we feel about our existence in general—depend in large part on our emotional, social, economic, and spiritual well-being. When we are saddened or afraid, when we are socially isolated or lonely, or when we are having financial difficulties, our lives are substantially less fulfilling. Many of the largest pharmaceutical companies now include quality-of-life measures to evaluate the effectiveness of their drugs that include emotional and psychological well-being, not just whether the patients are biologically better off.[7] There is a growing recognition that a life that is less emotionally, psychologically, and socially fulfilling is a less healthy life. And the World Health Organization (WHO) agrees. In its definition of health, WHO includes social, psychological, and physical well-being as important components.[8]

You may ask, "What is really new about this expanded, multidimensional view of health?[9] Isn't this just a repackaging of the old idea of holistic health?" This redefinition does indeed have some similarity with the concept of holistic health. Both share a core philosophy that health and illness are determined by many factors, not just the biological ones. But unlike the concept of holistic health, the new definition of health comprises dimensions that stem from, and are substantiated by, the highest standards of scientific research, research that has identified what I believe are six dimensions as the key determinants of health and longevity. It is this foundation in rigorous scientific methodology that sets the new definition of health apart from the philosophically similar idea of holistic health. In the former, notions about health and longevity are accepted or rejected based on the outcome of carefully designed studies. In the latter, scientific verification is less critical than are intuitive appeal and personal experience. Not that intuitive appeal and personal experience are unimportant. In fact, they are often the first phase of the scientific process. I personally enjoy a certain type of yoga, which I am convinced helped me heal an injury, but I have no scientific proof. However, I continue to practice it and believe that it helps me. By contrast, the dimensions of the new definition of health, and specific elements of each, have passed both the intuitive and the scientific tests. As you will see in subsequent chapters, however, intuitive and everyday beliefs about nonbiological dimensions of health are not always complete or accurate.

Research now conclusively shows that the more comprehensive, multidimensional, and expanded view of health described in this book is the most scientifically accurate one and will help us improve the understanding, treatment, and prevention of disease. Using the new definition, health is not defined as just the absence of disease, and sickness is not defined by deficiencies in any dimension of health. One of the criticisms of the WHO definition of health is that it defines health as the "complete state of physical, mental, and social well-being." Many scientists took issue with the word "complete" because it implied that if you are not experiencing 100 percent mental or social well-being (regardless of how it is defined), then you might be considered sick.[10] By such standards a majority of the world's population could be considered under the weather. My view of the new definition of health is that the dimensions should be viewed not diagnostically but, rather, prognostically. That is, deficiencies in any of the dimensions should not be labeled illness but should be looked at as (1) something that might increase or decrease overall quality and enjoyment of life and (2) something that might increase or decrease *risk* for actual disease. The first one is obvious. Having diminished emotional, social, psychological, or existential well-being is clearly not desirable and therefore can effect day-to-day experiences of life. The second one is less obvious and is the focus of this book. Dimensions of the new definition of health contain specific elements that are powerful determinants of longevity. Some of these elements increase risk for disease, while others protect health. As I pointed out earlier, having deficiencies in any of the elements does not represent illness, but it does represent warnings—signals that there is more we can do to optimize our health or avoid or recover from disease.

A theme of this new way of thinking about health is that of interconnectedness. Few illnesses that we suffer arise from biological processes alone, but instead from the interplay between our biological makeup, our social and physical environment, our thoughts and behavior, and our emotions.[11] This interplay determines in large part who gets sick, who stays well, and who recovers from, adapts to, or survives illnesses. Consider these facts about the six dimensions of health:

- *Each of the dimensions and its related elements is strongly and unequivocally linked to physical illness and longevity. Because of this, they all must now be considered crucial aspects of any definition of health.*

- *The dimensions are linked to one another, such that changes in one can precipitate changes in the others. In particular, changes in biological status (e.g., immune-system functioning) can be produced by changes in each of the other dimensions. Emotional, psychological, social, and spiritual well-being all have profound effects on the biological mechanisms that lead to illnesses.*
- *Health is often determined by an accumulation of risk and protective factors from across several of the dimensions.*

These three facts about the six dimensions of health are very important because they help us get away from a kind of either/or thinking about physical illnesses. That is, either my illness is due to a "real" biological cause, which means I can exclude nonbiological causes, or it's "all in my head" because an acceptable biological cause can't be found. In this book, you will learn that our thoughts, our relationships, and our emotions have real biological consequences and need to be part of our health profile as much as biological lab tests. In fact, biological and nonbiological risks for illness may work in combination. A great illustration of this combined effect is the research on ulcers, which is recounted by Dr. Robert Sapolsky in his classic book, *Why Zebras Don't Get Ulcers.* For decades, a good deal of research had shown a connection between stress and ulcers. Because of this kind of research, it became common knowledge among physicians and the public that ulcers could be caused by stress. Then, in the 1980s, it was discovered that a bacterium called *Helicobacter pylori,* or *H. pylori,* was associated with 85 percent to 100 percent of ulcers. The high prevalence of *H. pylori* in ulcer patients, coupled with the fact that treatment of the bacterium with antibiotics could cure the problem, led to a widespread rejection of the stress hypothesis for ulcers. A bacterium was in, and stress was out. Many biomedical scientists and physicians now say things like "We once thought ulcers were caused by stress, but now we know they are caused by bacteria" in the same tone as saying "We used to think the Earth was flat, now we know it is round."

But a closer look at the data provides a different picture. First, according to Sapolsky, up to 15 percent of people who get ulcers are *not infected with* H. pylori *at all.* More remarkably, although most people who have ulcers also have *H. pylori, only about 10 percent of people infected with* H. pylori *actually get an ulcer.* So if you get a certain type of ulcer, you are likely infected with *H. pylori.* But just because you are infected with *H. pylori* does not necessarily mean you will get an ulcer. The same can be said about stress. Just because you experience a great deal of stress does not mean you will get an ulcer—most people under intense stress don't get ulcers. But what is most fascinating about the ulcer research is that it now appears that ulcers are actually caused by *a combination of stress and H. pylori.* According to Sapolsky, in the presence of major life stressors, the addition of a small amount of bacteria inevitably leads to ulcers. On the other hand, in the presence of lots of bacteria, it takes only a small amount of stress to cause an ulcer. Of course, ulcers can be caused by many other things, not just stress and bacteria. But the point here is that factors within both the biological and nonbiological dimensions can combine to affect health outcomes.

THE EMERGENCE OF A NEW HEALTH SCIENCE

Three pivotal scientific trends, developing mainly within the last two decades, have fueled the expanded view of health, resulting in a new approach to health science. The first trend is the growth and increased methodological sophistication of research on nonbiological aspects of disease. Although there has long been an interest among social and behavioral scientists in factors such as stress, emotions, and social relationships and their connections to health, until recently this research has not met the highest scientific standards. Because of this, its acceptance within the medical establishment, such as in large medical centers and at the National Institutes of Health, was minimal. However, since the early 1980s there has been a proliferation of studies on the nonbiological elements of health that use the most rigorous research methods. These include (1) better measurement of concepts such as social relationships, stress, and depression, (2) longitudinal (i.e., long-term) studies to track the health of large populations over time, (3) randomized controlled trials to evaluate interventions, and (4) more sophisticated statistical analyses of study data.

The second trend has been more theoretical and philosophical in nature. It is the trend toward a clearer understanding and articulation of the manifold connections between our biological makeup, our psychological and emotional characteristics, and our social environments. The new view of health goes one step further than the concept that nonbiological factors are important to health. In the new view of health, these various dimensions are not truly separable. They are in fact inextricably connected, with each having the ability to influence, and be influenced by, the others. Nothing is *purely* social, biological, psychological, and so forth. No one element of health is self-contained—the status of each is multiply determined, even the status of our immune, cardiovascular, and neuroendocrine systems and the expression of our genes.

The third trend is a product of the first two: the rise of what is called *interdisciplinary research,* the sine qua non of the new health science.[12] Interdisciplinary research occurs when scientists from different disciplines, with expertise in the different aspects of health, combine their expertise to illuminate new connections between the various dimensions. Perhaps the most exciting type of interdisciplinary research has been between scientists trained in the behavioral and social sciences and those trained in the biological sciences. Scientists with backgrounds in psychology, sociology, anthropology, and public health are working with those trained in immunology, oncology, cardiology, and genetics. To those who are not health scientists, this may not seem like a major accomplishment, but the historic chasm between these fields has been oceanic. Social and behavioral scientists and biomedical scientists have largely ignored each other's research, even if they were all interested in the same health problem. When they did pay attention to one another, it usually consisted of one camp's criticizing the other as "soft science" and "irrelevant" to health, or "reductionistic" and "narrow." Very little real communication usually occurred across this chasm.

But all this is changing. The chasm is being bridged. Research in the health sciences is approaching a critical mass of social, behavioral, and biological scientists who are working together, making unprecedented discoveries, and changing the terrain of health science. This interdisciplinary research is the heart and soul of the new health science. What they are finding is intriguing and often surprising. This book chronicles some of the most important and fascinating findings of these studies.

Each chapter of this book describes a small sample of some of the most exciting, innovative, and cutting-edge research that provides clear illustrations of the six dimensions of the new definition of health and their connections with one another. I say "small sample" of research because the explosion of science in this field makes it impossible to describe everything. So I picked topics within each dimension of health that I thought were exemplary. In this book, you will learn that your physical health and longevity can be affected by:

- your expectations about the future
- how you explain things that happened in your past
- your friendships and social ties
- your education and income
- the degree of control you have in your work
- traumatic experiences you have never disclosed
- your ability to find meaning in negative life experiences
- your experience of three key emotions

These and other topics featured in this book do more than shed light on and provide a scientific basis for the new definition of health. The studies I will describe to you often confirm or refute proverbial wisdom about what is healthy and what is not. Is it always good to look on the bright side? Does every cloud have a silver lining? Is venting emotions the best way to handle pent-up distress? Are the health benefits of marriage as good for the goose as for the gander? Is it best always to be realistic, to see things clearly and accurately? The answers to these and other questions might surprise you.

Finally, but of equal importance, is the fact that this book is about more than science. It is also about people. People whose lives put a human face on the amazing discoveries I will share. Some of these people you might recognize, such as entrepreneur Wally "Famous" Amos, *Parade* magazine editor Walter Anderson, author and poet Maya Angelou, television professional Linda Ellerbee, Duke University basketball coach Michael Krzyzewski, and former Atlanta mayor and U.S. ambassador to the United Nations Andrew Young. Others may be unfamiliar to you, but they have equally compelling stories. Their lives exemplify how the new dimensions of health can be used to achieve a high quality of life and overall emotional and physical well-being.

PART I

THOUGHTS AND ACTIONS 1: EXPECTATIONS, EXPLANATIONS AND BELIEFS

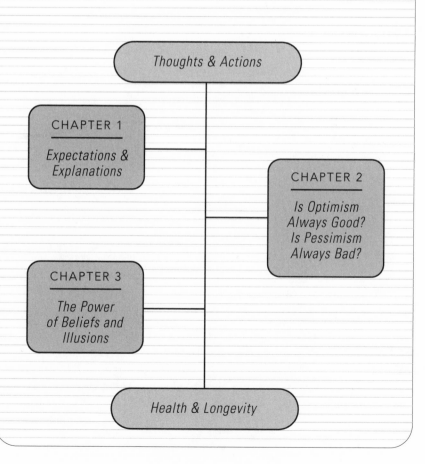

Thoughts & Actions

CHAPTER 1

Expectations & Explanations

CHAPTER 2

Is Optimism Always Good? Is Pessimism Always Bad?

CHAPTER 3

The Power of Beliefs and Illusions

Health & Longevity

The facts of our lives are not as important as our attitudes toward them.

—Viktor Frankl

To say that our thoughts and actions can affect our longevity is to say something old, but also relatively new. The old is that what we do—our actions and behavior—is health enhancing or health damaging. Eating a proper diet, getting regular exercise, avoiding cigarettes, ingesting moderate amounts of alcohol—these all have now been enshrined on the public's to-do list. So I will not belabor those points. I would much rather discuss the much less widely recognized determinant of our health—our thoughts.

More specifically, this part of the book deals with how our thoughts about the past and the future, and even our beliefs about ourselves, can influence our actions, affect our emotional well-being, and even determine how long we live. No one knows this better than Art Berg.

Lying on a highway forty miles north of Las Vegas in the expanse between California and Utah, waiting for an ambulance in the cold darkness that four-thirty brings to desert mornings is not when most people would have an optimistic thought. But Art Berg isn't most people. "At first I thought I had hit a new level of bad. But the thought that came to my mind quickly after that was 'This may give me an unprecedented opportunity as well.'" Berg, just twenty-one years old in 1983, had been en route to Utah from California in December for a holiday visit. This was a trip Berg had made countless times before, to visit his fiancée, a young woman whom he'd dated since he was sixteen. He drove the first leg for almost nine hours before turning the wheel over to his friend John, who was supposedly rested. John fell asleep while driving, and the car hit a concrete barrier and rolled four or five times down the side of the road. Berg had his seat belt on, but his seat was reclined. His car had a passive restraint system—just a shoulder harness, no lap belt. Berg apparently slipped out of the seat belt and went through the back window. He landed on his forehead and snapped his neck. He spent four months in the hospital and remains a quadriplegic. But Art Berg entered the hospital with a perspective on the world that inoculated him from the natural emotional fallout of such tragedies. He had a belief in himself and a way of interpreting the events of his life that has helped catapult him to own several successful businesses, to author two inspiring books, to become a sought-after motivational speaker, to set world records as an athlete, and to even earn a Super Bowl ring.[1]

Few of us could view the situation that Berg found himself in as possibly presenting an "unprecedented opportunity," the way Berg did. Berg himself admits that he can be a little extreme in his attempts to interpret ostensibly negative events in their most favorable light. But as extraordinary and surprising as Berg's attitude was following his accident, his mind-set is nevertheless a vivid exemplar of the power of our thoughts to transform our reactions to even the most life-altering events. In Berg's case his unwaveringly positive expectations about his future—his enormous optimism—helped him triumph over tragedy.

There is perhaps no better example of the importance of the psychological dimension of the new definition of health than the research on expectations, explanations, and beliefs. Our expectations, explanations, and beliefs about the world and about ourselves function in many respects like antibodies of the mind. In the way that antibodies destroy cells and organisms that threaten the well-being of the body, so do our expectations, explanations, and beliefs provide us with some defense against negative life experiences that could harm us emotionally and physically.

With our expectations we can construct a vision of an uncertain future that is filled with positive anticipation and hope. We can look back and explain the past in ways that are less of an affront to our well-being. Indeed, we can see the past through rose-colored glasses— not denying that things happened but seeing them in ways that are less emotionally jarring, less psychologically scarring. We can develop beliefs about ourselves and our abilities that motivate our actions in ways that enhance our health and help us achieve our goals.

The way we formulate expectations about the future and construct explanations of the past establishes our view of the world. Do you tend to look on the bright side of things? Is the glass half empty or half full? Does every cloud have a silver lining? We often use the answers to these clichéd questions to characterize ourselves or others as optimists or pessimists, but these simple assessments overlook what is at the core of being optimistic or pessimistic. The essence of optimism and pessimism is much more profound and meaningful than clichés may suggest.

Scientists who study optimism are really exploring two essential processes: (1) how we *expect* events to turn out in the future and, paradoxically, (2) how we *explain* events that have already occurred. What we expect of the uncertain future and how we explain and interpret the past to ourselves and others determines whether we are optimists or pessimists. Optimists expect and project a positive image of the uncertain future and explain and interpret the past in a favorable light. They believe they can overcome any apparent adversity and achieve success. Pessimists lack a belief in their abilities and characterize their expectations and explanations in negative terms. At the root of such expectations and explanations is a core belief system about ourselves and the world.

Our explanations, expectations, and beliefs about the past and future are not benign mental constructions. In the way that reins serve to restrain a horse's motion, our thoughts can hamper—or enhance—our motivation and our inclination to take action. Becoming a quadriplegic certainly had the potential to constrain Art Berg's life. Yet he was able to persevere, in large part because of his belief that regardless of how bad his circumstances were, things could and would be better. Such thoughts, he says, kept him from sinking into hopelessness and despair. Indeed, our expectations of the future and our explanations for the past can leave us calm and hopeful, with feelings of well-being and confidence about the future. They can also leave us distressed, anxious, or downhearted. Perhaps more remarkable, however, is that our expectations, explanations, and beliefs also conspire to have profound effects on our physical health and longevity.

Chapter 1

Expectations and Explanations

DO YOU EXPECT THE BEST OR THE WORST?

Uncertainty is unsettling. We are most comfortable when we think we know what's going to happen, when it's going to happen, and what the short- and long-term consequences will be. But the outcome of most events in our lives is essentially unknown.

Will you do well in your new job? Will problems in your relationship be resolved successfully? Will your wedding go as planned? Will you get to the airport on time? Will you reach certain goals? The outcome of many of the challenges that confront us is unpredictable and frequently beyond our control. Research has shown that unpredictability and uncontrollability are extremely stressful and that to cope with the stress of uncertain outcomes, we often create imaginary futures in our minds. In essence, we fill the void of uncertainty with our personal expectations and our idiosyncratic hypotheses about how things will turn out. The expectations we generate make the future appear a little less in doubt and help us feel more in control.

The expectations we place on the uncertain future are not inconsequential. Indeed, they have a powerful influence over what we will actually experience, and they even have far-reaching health consequences. For example, when inactive or fake drugs produce a biological effect—called the placebo effect—that experience is fueled by expectations. We expect the pill to work; it does.

The field of psychology has a long history of research on how expectations about the future drive so much of our behavior. The nature of our expectations—whether we anticipate that outcomes will be desirable or disagreeable—determines whether we are optimists or pessimists. Optimists and pessimists can be differentiated by their responses to questions such as these:

> In uncertain times do you expect the best?
> Do you believe that if something can go wrong, it will?
> Do you rarely expect things to go your way?

The above questions are from one of the most widely used evaluations of expectations, the Life Orientation Test, one of several tools scientists use to determine if a person is an optimist or a pessimist.

Psychologists Drs. Michael Scheier and Charles Carver, collaborators for more than twenty years, developed the Life Orientation Test to determine whether an individual tends to expect future outcomes to be good or bad. Scheier and Carver believe that all of us have characteristic ways of envisioning the future. Some of us are optimists, primarily inclined to expect good outcomes across a variety of situations. Some of us are pessimists, inclined to expect bad outcomes across a variety of situations. In fact, some people are called *dispositionally* optimistic or *dispositionally* pessimistic because they are so consistent in their positive or negative expectations. Dispositional optimists and dispositional pessimists seem to operate in different emotional worlds. The optimists experience generally positive moods characterized by happiness, excitement, and interest. The pessimists experience more negative mood states, often accompanied by distress and anxiety.[1]

Most of us fall somewhere between the two dispositions, neither routinely expecting the best nor routinely expecting the worst. No one is either completely optimistic or pessimistic all the time. But when our lives are derailed by trauma or an unexpected and dramatic event, each leopard shows his spots. That's when our optimistic or pessimistic tendencies are revealed.

The Leopard Shows His Spots

Research indicates that the benefits of optimism and the liabilities of pessimism are most clearly seen during the worst of times. When life is going well, it is really difficult to distinguish optimists from pessimists. It is when life presents its inevitable challenges to our well-being that our optimistic or pessimistic dispositions kick in. At times of adversity or during stressful life circumstances, it is indeed true that optimism is beneficial. But what has surprised researchers is just how destructive pessimism is. Optimism, it turns out, protects health. Pessimism, on the other hand, increases vulnerability to emotional and physical dysfunction.

Consider the varying effects optimistic and pessimistic women experience in response to the complex factors of pregnancy and childbirth. For many women the transition from pregnancy to motherhood is laden with stressful issues: worries about adjusting to the new role of parent, concerns about changes in marital and familial relationships, added financial pressures, changes in their bodies and body image. Studies show that pregnancy, especially in the latter stages, can take an emotional toll. Not surprisingly, many women experience symptoms of depression following the birth of a child. When those symptoms persist for weeks and months after delivery, the new mother is said to suffer from postpartum depression. A few women, such as Andrea Yates, who admitted to drowning her five young sons in the summer of 2001, suffer so severely that their condition can be considered postpartum psychosis.[2]

One of the first studies of dispositional optimism revealed that women who are optimistic are resistant to developing symptoms of postpartum depression.[3] The study found that women high in dispositional optimism had lower levels of depression prior to delivery and, more important, suffered significantly fewer symptoms of depression during the three weeks after childbirth. In contrast, dispositionally pessimistic women were significantly more likely to experience postpartum depression. These findings persisted even after statistically controlling for the level of depression prior to delivery.

In another study, researchers at the University of Pittsburgh explored the effects of dispositional pessimism on depressive symptoms over a three-year period in 460 premenopausal women, ages forty-two to fifty.[4] The study found that one of the strongest predictors of depressive symptoms was dispositional pessimism. The highest levels of depressive symptoms occurred in those women who were pessimistic *and* who had experienced chronic stress during the study period.

In addition to influencing emotional well-being and the ability to cope, dispositional optimism and pessimism affect a host of other health outcomes. Here are some surprising ones:

Dispositional optimism predicts recovery from bypass surgery. In two long-term studies of patients undergoing coronary artery bypass surgery, Scheier and his associates assessed whether dispositional optimism is related to a better recovery.[5] In the first study the patients with high dispositional optimism recovered more rapidly than did the dispositionally pessimistic patients. The differences between the optimists and pessimists emerged as early as one week after surgery and were still apparent as long as six months later. During the first week after surgery, the optimists were better able to sit up in bed and walk around the room than were the pessimists. Hospital staff, not knowing how the patients were classified, rated the optimists as showing a more favorable physical recovery than the pessimists. Six months after surgery the optimists were more likely than the pessimists to have returned to work full time, to have resumed vigorous physical exercise, and to have reestablished a normal level of functioning in many aspects of their lives.

Dispositional pessimism predicts rehospitalization after bypass surgery. The second Scheier study assessed optimism and pessimism in 309 patients scheduled for coronary bypass surgery. Six months after surgery patients who were pessimistic were more likely to require rehospitalization for complications from surgery or other heart problems.[6]

Dispositional pessimism predicts mortality in cancer patients. A remarkable long-term study explored the negative effects of pessimism in more than 200 patients with metastasized or recurrent cancer who were receiving radiation treatment.[7] Assessments of dispositional optimism and pessimism were made when the patients entered the study and again four months and eight months later. A total of 70 patients died during the study, and there was a striking association between dispositional pessimism and death. Patients thirty to fifty-nine years old who scored high in pessimism were significantly less likely to survive to the eight-month follow-up assessment. Optimism, surprisingly, was unrelated to survival. Intuitively, you might think that if pessimism was associated with a high risk of death, optimism would predict greater survival. But an interesting and perhaps important observation made by the researchers is that optimism and pessimism may not be simply opposite sides of the same coin: A person can lack optimism but not be pessimistic, and vice versa. It is almost as if we have within our personalities separate optimism and pessimism dimensions that can operate quite independently. We can be high on one dimension but not necessarily very low on the other. Thus, it can be possible, as in this study of cancer survival, that one dimension can be related to longevity, but not the other.

Dispositional optimism and pessimism predict immune status in law students. A study by researchers at UCLA investigated the effects of optimism on mood and the immune system in 50 law-school students during their first few months of school (a time of very high stress). This study examined both dispositional optimism and situational optimism. Situational optimism was defined as an optimistic expectation specifically for success in law school. Baseline assessments of optimism and blood samples for immune-system measures were obtained before the first semester began or on the first day of classes. Follow-up assessments were conducted during weeks eight and nine of the semester (midsemester). Both optimism and pessimism were strong predictors of both mood and immunological functioning. Optimism, especially situational optimism, was strongly related to higher levels of crucial indicators of healthy immune function, such as T cells, B cells, and NK cells.[8]

Dispositional optimism and pessimism predict blood pressure. In a project designed to examine the role of optimism on blood pressure throughout the day, 50 women and 50 men wore ambulatory blood-pressure monitors for three days (two workdays and one nonworkday). In this study, when the blood-pressure cuff inflated and a reading was recorded by the monitor, the participant made a diary entry to rate current mood state (e.g., stressed, angry, happy). Blood pressure throughout the day was higher in the more pessimistic than the optimistic participants.[9]

You can see that expectations of the future can influence our emotional and physical well-being. However, there is another way of thinking about optimism that is not at all focused on the future but surprisingly has to do with our thoughts about the past. In the next section I will explore how our thoughts about, and especially our interpretations of, events gone by are also important to our health.

EXPLAINING AND INTERPRETING THE PAST

How do you explain bad events? What implications do you draw from the explanations that you develop? Do you think that whatever happened is your fault? Do you think that such things happen all the time? Do you think that things will never change? Your answers to questions like these are indications of how you explain good and bad events, and are another way of determining whether you are an optimist or a pessimist. Not long ago my wife and I came face-to-face with the effects of explanations.

Shortly after I decided to leave my position as associate director of the National Institutes of Health for a professorship at Harvard University, Elizabeth and I embarked on a housing search in New England. We once again discovered why moving is considered one of life's top stressors. We wanted to live in either the greater Boston area or nearby Providence, Rhode Island, where I'd lived in the early 1980s while completing my clinical-psychology internship at Brown University. We ruled out Boston in short order as prohibitively expensive and focused on Providence. But the Providence real-estate market was also exploding, albeit less so than Boston's. Sellers were getting either full-price offers or even more through competitive bidding. To make matters worse, there were simply few houses on the market in our price range. Having to travel to Providence for housing searches while continuing to work in Bethesda, Maryland, complicated matters even more.

We thought about renting, taking our time to look around, and waiting out the market. But we really didn't want to make two short-term moves, and I wanted the housing issue wrapped up before starting a new job. Plus, we had a book to write (this one), and we needed to have all our research materials, files, and computers out of boxes. So as the time of relocating approached, the pressure to find a place intensified.

After a number of visits to Providence, we were still coming up empty. Houses either were too expensive, needed too many renovations, or were in an undesired location. Now it was getting close to crunch time, when we absolutely *had* to find something. We were fortunate to visit Providence on a day when a hundred-year-old Victorian was coming on the market. Although it cost more than we preferred to spend, it was in a great location. Our real-estate agent was being very patient but encouraged us to act fast if we wanted the house, because it would not last long in that market. So after viewing the house twice that day, we phoned in a full-price offer from the airport on our way home. The offer was accepted, and we soon signed a contract. Then psychological disaster struck.

The basic problem was that this was not how we would typically buy a house. Our style would have been to see the house a number of times, mull over the details, and allow the idea of buying it to settle in our minds. This process could take a few days or weeks. But we bought in a day! In fact, we bought only about eight hours after first laying eyes on the place. It wasn't long before a serious case of buyer's remorse kicked in. All of the house's shortcomings became salient. The rooms were too small. Not enough space for our stuff. Too many renovations needed. Too many floors. Too narrow. Not enough sunlight. Must be something better out there.

Too late. We were in too deep. We had to go through with the purchase.

Although it turned out that the house could in fact contain all our stuff, most of our other concerns were validated after we moved in. Plus, we discovered new problems to make things even more challenging. But the point of this story is not the house per se but the widely divergent emotional reactions that Elizabeth and I experienced after moving in. In a few months I came to accept the fact that this house is where we live, and I actually developed a level of appreciation for our home and thought we were in a great location.

Elizabeth's reaction was totally different. The house for her became a source of endless distress and a constant trigger for a depressed mood. Her dissatisfaction with the house grew daily. The more she lived in it, the more she found to dislike. One couple owned both of the houses on either side of us, and they were doing major renovations to them at once. The constant noise, dust, and disruption nearly drove Elizabeth mad. The way she felt about the house clouded every aspect of her life in Providence for more than a year. She was nearly overwhelmed with what she saw as a terrible mistake that we would not get out of without having to endure unimaginable hardship.

Why such disparate reactions to the same situation? Our emotional reactions about the house stemmed from our divergent perceptions about it. We each attributed our situation and its outcome to completely different factors, which resulted in different emotional consequences.

Our opposing reactions would come as no surprise to Dr. Martin Seligman, a world-renowned scientist and former president of the American Psychological Association. In 1975 Seligman encountered a major challenge in his research—one that threatened his principal claim to fame. However, his solution to the problem revolutionized scientific views on the effects of thoughts on mood, achievements, and health. It also provided an explanation for, and a way out of, the emotional tangle created by our housing dilemma. More on that in a bit.

In the late 1960s and early 1970s, Seligman and his colleagues at the University of Pennsylvania formulated a theory of depression, which they called "learned helplessness." The theory grew out of years of laboratory research on the behavioral and health effects of uncontrollable stress. Animals exposed to uncontrollable stress, like a mild electric shock, when given a way to escape (say, by moving to a safe area), would always take it. When the escape option was removed, they would for a time continue to search for ways to escape. As time passed, however, the animals just gave up. They would stop trying to escape the mild shock and simply endure it. Intriguingly, when they were again given the opportunity to escape, they would refuse to take it. Even though they were returned to conditions in which their behavior could relieve the stress, they remained passive. Seligman called this passivity "learned helplessness" because the animals acquired the "knowledge" that their efforts would be futile, even if this were not true. They became helpless and passive after repeatedly experiencing uncontrollable circumstances. When their environment was again controllable, their passivity remained. They had learned to be helpless.

Seligman's research team discovered that many humans also showed helplessness when confronted with chronic, uncontrollable stressful conditions in the laboratory. Like the animals in the earlier studies, people also become passive and are prone to give up when things seem out of their control. In essence, they explain to themselves, "Nothing I do makes any difference." Even when conditions change and the situation becomes more controllable, their passivity remains. They have learned to be helpless. Many people, when exposed to conditions they perceive as uncontrollable, develop depressive moods and other symptoms of clinical depression. Seligman holds that depression is due in large part to people's perceptions that events are not in their control. A perceived lack of control leads to depressed mood, passivity, and giving up—in other words—to helplessness.

In general, the theory worked quite well in accounting for a number of research findings on depression. However, a problem with the theory was that exposure to uncontrollable conditions did not *always* lead to helplessness and depression. Seligman's research revealed that many people who experience unrelenting stress, major setbacks and failure, or potentially crushing disappointment simply do not become depressed and do not give up. I asked him to explain why this happens. He said, "People with an optimistic explanatory style resist helplessness more easily than people with a pessimistic style. For an optimist, a setback is temporary, local, and controllable."[10] Plainly put, optimists view the past in its best possible light.

Seligman and his associates found that *explanatory style*—your routine way of explaining bad events (not the bad events alone) can lead you down the path to helplessness and depression. We explain an event in our lives, especially a bad event, along three dimensions having to do with whether we take *personal* responsibility for it, the degree of *permanence* we feel the event has, and the level of *pervasiveness* of its effects. When an untoward event occurs, we use these dimensions to help us come to terms with why it occurred and its implications. These dimensions are the keys to whether the explanation is in the realm of optimism or falls on the side of pessimism.[11] *Personal* explanations for bad events determine our degree of self-blame. Does the bad event reflect some personal flaw, or is it due to the circumstances? *Permanence* addresses our thoughts about how long we expect the circumstances created by the bad event to last. Will it be chronic or short-lived? *Pervasiveness* deals with how all-consuming the event will be. Is it going to affect how you experience other things in your life, or will your feelings be confined to this narrow incident?

Some of us use all three dimensions to explain bad events, and research suggests that the more our thinking about such events reflects self-blame, permanence, or pervasiveness, the more depression we will experience. If you say, "It's my fault," if you feel the event is permanent, and if you feel it is pervasive, then it seems uncontrollable.

Perceptions of uncontrollability lead to feelings of helplessness. Feelings of helplessness lead to depression.

So what does explanatory style have to do with optimism or pessimism? Quite simply, pessimists explain bad events in ways that lead to the anticipation and expectation of more negative events in the future. Optimists, on the other hand, explain good events in ways that lead to the anticipation and expectation of more good events in the future. If you feel that a negative or positive circumstance occurred because of you *personally,* that it is *permanent,* and that it is *pervasive,* it seems to you that things are unlikely to change. That's fine if the event is positive—you don't want those things to change. But for untoward life events, change is exactly what you want.

When unfortunate situations occur, optimists are more likely to say, "It's the circumstances," "It's temporary," and "It doesn't affect anything else in my life." Pessimists are more likely to say, "It's because of me" or "It's my fault," "It's permanent or long lasting," and "This messes up everything." But, ironically, when really good things happen, optimists and pessimists reverse strategies, with optimists thinking, "It must be because of me," "It is going to last a long time," and "It positively affects everything else." Pessimists, on the other hand, adopt the perspective that positive situations are unique to the circumstances ("It was luck. I had nothing to do with it"), are not permanent ("Enjoy it while it lasts; it won't last long"), and affect nothing else ("It makes no difference"). Given these habitual ways of explaining the past, it's not surprising that pessimists have higher rates of depression and negative emotional states than do optimists.

The different emotional reactions that Elizabeth and I had to our housing situation were associated with our underlying explanatory styles. My thinking was that the decision to buy the house did not reflect negatively on me—it was just that we needed a house quickly ("It's not me, it's the circumstances"). I saw the situation as temporary, since we would probably move in a few years ("It's not permanent"). And I believed that the house certainly would not affect my overall enjoyment of life ("It's not pervasive"). Elizabeth had the opposite explanations. To her, purchasing the house reflected badly on her and her influence on our decision making ("It's me"). She did not experience living in the house as a temporary negative event. For her it was going to last a long time ("It's permanent"). And it would negatively affect her enjoyment of other activities in Providence ("It's pervasive").

During the first year in the house, we had many, and sometimes very tense, discussions about the house and our polarized feelings about it. About a year after we moved, some particularly disruptive renovations next door caused Elizabeth to be especially distraught about the house. It was causing her great emotional pain, and she repeatedly questioned why we ended up with such a bad purchase. In response, I explained, "Honey, it wasn't a bad decision, because we needed to quickly establish a home to begin writing the book, and there were very few on the market" (it's the circumstances). We had spent five years renting in Maryland, and there were advantages to doing so, but we had taken a beating on our taxes, so I reminded her, "Plus, we needed to buy a house quickly for tax purposes" (it's the circumstances). "But you're right, it may not be the perfect house for us, and we should definitely look for a new one when the book is done" (it's temporary).

After hearing and discussing this, Elizabeth experienced a positive change in her mood. She said that she felt less trapped because we developed a plan to change (and in her mind improve) our situation. She felt less responsible for the house purchase and accepted that we had been led more by the circumstances than by her will. Not that she can't be persuasive when she wants, but she realized that she hadn't forced this move: It was not her fault. We did not have to live there forever. She could go about enjoying other aspects of her life without feelings of dread.

What's Your Explanation?
Health-Enhancing Explanatory Styles

Researchers measure explanatory style with one of two tools: (1) the Attributional Style Questionnaire (ASQ) and (2) the Content Analysis of Verbatim Explanation (CAVE) technique. The ASQ is used widely to determine people's explanatory style by having them speculate on the causes of hypothetical bad events. On a scale of one to seven, participants rate the degree to which they view an event as (a) personally caused, (b) permanent, and (c) pervasive.

Dr. Christopher Peterson, a close colleague of Seligman's, developed CAVE to determine a person's explanatory style using what the person wrote or had been quoted as saying. Employing documents such as letters, autobiographies, or even newspaper or magazine interviews, researchers can determine via the CAVE technique whether someone is optimistic or pessimistic based on how he or she explained some bad event. Researchers rate the degree to which the bad event is viewed as personally caused, permanent, or pervasive. CAVE is so powerful that it has been used to predict everything from presidential and senatorial elections to the outcomes of professional basketball and baseball games.[12] Even documents related to deceased persons can be used. For example, the CAVE method was used to "predict" the outcome of presidential elections between 1900 and 1984.[13] Researchers analyzed the nomination speeches of both Democratic and Republican candidates. Those candidates whose speeches were judged as having more pessimistic statements actually had lost nine out of ten elections, even though the people assessing the speeches did not know whose they were. Pessimism even predicted upset victories, such as

Truman over Dewey in 1948, Kennedy over Nixon in 1960, and Reagan over Carter in 1980.

Using the ASQ and the CAVE technique, scientists now know that explanatory style also affects health and emotional well-being in the following ways:

Explanatory style is associated with depression in adults. Persons with diagnosed clinical depression tend to use a pessimistic explanatory style for bad events, and the more severe the depression, the more pessimistic the explanatory style.[14] The depression lessens when therapy produces changes in the explanatory style from pessimistic to more optimistic.[15]

Explanatory style predicts depression in children. In a long-term study, Seligman and colleagues found that children's explanatory style is a strong predictor of subsequent depression.[16] One study assessed explanatory style and negative life events (e.g., parents' divorce and arguments, grandparent's death, social rejection) in third-graders over a five-year period. Although negative life events emerged as the strongest predictor of depression, a pessimistic explanatory style was associated with depression as children became older. The researchers speculated that "early in life children's explanatory style may still be under development, and negative life events, not cognitions, predict periods of depression. As children grow older and their cognitive capabilities increase, their explanatory style becomes more stable and appears to play a stronger role in the development of depressive symptoms."

Several long-term studies have now linked explanatory style with physical health and longevity, including the ones that follow.

Explanatory style predicts illness. The CAVE technique was applied to transcripts of responses to open-ended questions answered by 99 graduates of Harvard University classes of 1942–44, who were then about twenty-five years old. The graduates had regular physical examinations from ages thirty to sixty. *Those with a pessimistic explanatory style for bad events were significantly more likely to suffer* *diagnosable physical illnesses over this period of time than those with more optimistic styles.*[17] In a study by Dr. Laura Kubzansky at the Harvard School of Public Health, the relationship between optimistic and pessimistic explanatory styles and heart disease was explored in 1,300 older men, who were followed for ten years as part of the Veterans Administration Normative Aging Study.[18] Although the men were healthy at the start of the study, over the ensuing ten years, 162 of them developed some form of heart disease, including 71 with nonfatal heart attacks, 60 with angina pectoris, and 31 who died of heart ailments. Participants who were higher in pessimism were significantly more likely to develop some form of heart disease compared to those higher in optimism. In fact, *men high in optimism were less than half as likely to develop heart disease than were the more pessimistic men. For every increase in optimism scores, there was a corresponding decrease in risk for heart disease.*

Explanatory style predicts longevity. The CAVE approach was also used to analyze explanatory style in transcripts of responses to open-ended questions provided by participants in the Terman Life-Cycle Study, a study of gifted and talented preteen boys and girls. The study began in the 1920s, when the more than 1,500 healthy children were in public school, with follow-up assessments every five to ten years. *The tendency to catastrophize negative life events—that is, attribute them to pervasive causes—was a powerful predictor of early death over the course of the next thirty years, especially among the males. Catastrophizing most strongly predicted deaths from accidents and violence.*[19] A study of more than 800 patients at the Mayo Clinic examined the association between explanatory-style assessments taken in the mid-1960s and all causes of death thirty years later. Pessimistic explanatory style predicted death in a surprisingly systematic way: *For every ten-point increase in pessimism, there was a corresponding 19 percent increase in risk for death.*[20]

Explanatory style predicts immune status. Exactly how a pessimistic explanatory style leads to illness and death is unclear, but scientists are beginning to explore the biology of such attributions. For example, among the elderly, a pessimistic explanatory style has been associated with poor cell-mediated immunity.[21] Among HIV-positive persons those who attributed negative events to some aspect of themselves (e.g., "I lost a couple of friends because I am HIV positive") had a more rapid decline in helper T cells (CD4 cells) over the course of eighteen months than did those who made fewer such attributions.[22]

Explanatory style predicts pulmonary function. Harvard researchers have recently explored the connection between explanatory style and pulmonary (lung) function in older individuals.[23] The reason this was of interest is that pulmonary function declines with advancing age and some individuals are more "rapid decliners" than others. An accelerated decline in pulmonary function increases susceptibility to chronic obstructive pulmonary disease, the fourth leading cause of death in the United States.[24] To determine if optimism was associated with pulmonary function, researchers measured explanatory style and pulmonary functioning of 670 older men as part of the Veterans Administration Normative Aging Study. Participants had at least three pulmonary exams over the course of eight years. *It was discovered that men with a more optimistic explanatory style had significantly higher levels of pulmonary functioning than did the more pessimistic men. It was also found that the decline in pulmonary function over the course of eight years was substantially less in optimistic than in pessimistic men.*

The effects of optimism, whether defined as our expectations for the future or as our explanations of the past, sound almost too good to be true. It has a positive effect on our mood, our biological status, and our life span. But what are its limits? Is optimism always good and pessimism always bad? In the next chapter I will examine these questions.

Chapter 2

Is Optimism Always Good?
Is Pessimism Always Bad?

Seems as if there is nothing better for good health and longevity than optimism, but can there be a downside to such a rosy worldview? Are optimists out of touch with reality? Could habitually having a positive outlook set you up for a fall if things turn out different than expected? Could the expectation of positive outcomes lead to passivity?

These are some of the typical questions raised about optimism by scientists and the general population. As remarkable as it may sound, the answer to each is no. There is very little research to suggest that optimism, either dispositional or explanatory type, has a significant downside. It could be argued that optimists deny the ugly underbelly of life and live in a fantasy world. Is it possible that optimists could construct such a positive and upbeat view of life that reality can only fail to live up to it? When it does fail, optimists could be in for a fall. But fortunately, this does not seem to be the case.

- *Are optimists in denial, turning a blind eye to unpleasant realities?*
  Charles Carver of the University of Miami explained to me, "Our research with cancer patients makes it clear that optimists are more likely to accept the reality of the cancer diagnosis and move more directly to trying to decide what to do about the changes that the diagnosis brings with it." As you will read later in the next chapter, although they accept the reality of their diagnoses, many optimistic cancer patients paint a rosier picture than do pessimists about aspects of their cancer that are more speculative, like their prognosis. And they may even live longer as a consequence.

- *Are optimists more vulnerable to the negative impact of a traumatic event, since it contradicts their worldview?* No, according to Charles Carver. He says that he "has seen no evidence that optimists are more vulnerable to the negative impact of traumatic events. It is possible that if an event is extreme enough, optimists might be shattered. But I think an event that extreme would have an even larger impact on the pessimist, who is less resilient. As a recent test of this, we assessed emotional responses of people a week after the attack on the World Trade Center. Although our results are preliminary, optimism appears to be related to slightly lower reports of depression."
- *Do optimists exhibit less healthy behavior, since they believe that their health will be fine anyway?* According to Carver there is evidence that optimists exhibit *more* healthy behavior. He says that they seem to believe that the world is an orderly place, but also that you have to stay in the game in order to win. To optimists this means doing whatever they can to ensure their health.
- *When ill, do optimists and pessimists differ on how well they take care of themselves?* Carver told me that in a study he and Michael Scheier conducted of men who had bypass surgery, optimists were more likely to be taking good care of themselves months later, such as being more apt to enroll in rehabilitation programs. He explained, "It is as if the optimists, when ill, have a stronger belief that through their actions they will get better faster. You might say it is the opposite of fatalism."
- *Do optimists blame others?* One concern about the explanatory-style version of optimism is that people might take the "personal" dimension to the extreme and not accept responsibility for their actions when they should. The "it's not me, it's the circumstances" response could devolve into "it's *always* someone or something else's fault." I asked optimism expert Christopher Peterson about this, and he says there is really little fear of this happening. "There is a strong tendency for people to look inwardly, at themselves, when affixing blame or responsibility for negative events. People who are depressed do this even more than others. So the realistic goal is to help people move away from the extremes of self-blame and to at least be neutral about assigning responsibility. Although there are always exceptions, most of us are unlikely to go from extreme self-blame to consistently blaming others for our woes," he explained.[1] Both

Peterson and Seligman argue that becoming optimistic is actually about seeing the world more accurately—taking responsibility for a bad event when appropriate and trying to correct the situation, but refraining from taking personal blame when the problem resides elsewhere.[2] Pessimists seem to have a bias toward blaming themselves too much, and a more accurate assignment of responsibility following bad events would be a potentially health-enhancing improvement.

On the whole it appears that there are minimal downsides to optimism. Being optimistic is a lot like wearing rose-colored glasses—they make reality less jarring and glaring without concealing it. In addition, when stress arises, optimist and pessimist use different strategies for coping with it. By "coping" I refer to things people do to reduce the negative emotional, psychological, and physical fallout of stressful life circumstances and daily travails.[3] Research has examined how dispositional optimists and pessimists cope with extreme life difficulties such as cancer, chronic work stress, heart disease and coronary bypass surgery, academic pressures, failed in vitro fertilization, and AIDS.[4] In general, this is how an optimist copes with stressful life events:[5]

- *Focuses on the problem at hand.* When faced with challenges that are potentially controllable (i.e., where his or her actions can make a difference) an optimist focuses on the problem and how to solve it.
- *Accentuates the positive.* An optimist uses positive reframing to interpret the problem in a positive but realistic light.
- *Lightens up a bit.* An optimist uses humor to regulate negative emotions and often tries to find the benefits that might result from life's adversities.
- *Accepts things for what they are.* When challenges cannot be controlled, an optimist accepts the situation.

A pessimist, on the other hand, copes quite differently with stress.

- *Focuses on the negative.* A pessimist tends to dwell more on the emotional distress and the negative aspects of a situation, rather than on solving the problem at hand.

- *Shuns perseverance.* A pessimist gives up more readily in the face of difficulties and often disengages from problem-solving efforts.
- *Avoids reality.* A pessimist makes greater use of overt denial (i.e., refusing to accept the reality of the circumstances), particularly when the challenges appear to be uncontrollable.

Young Mattie Stepanek effectively uses optimistic coping. Born in 1990, he copes daily with a rare form of muscular dystrophy that threatens his life every day. Mattie confronts death by living exuberantly. Confident in his abilities, he says, "I live with my disease every day. Everything is a chore, even getting out of bed. I have to do many things, but it's okay. I can cope with it." The disease that restricts him to a wheelchair and requires him to use a ventilator took the lives of his three siblings and afflicts his mother. However, Stepanek does not see himself as burdened and does not look outside himself for answers to his situation: "Better me than a little baby or a child that's already got a lot of stress in their minds. So I'll think, 'Why me?' And I'll think again, and I'll say, 'Why not me?' Better me than someone else." He looks only for ways to contribute to society and spread peace just like his favorite peacemaker, President Jimmy Carter, who wrote the foreword to one of Mattie's books of poetry. Stepanek has been writing poetry, the vehicle for his peacemaking efforts, since he was three years old. His books made the *New York Times* bestseller list, and the Muscular Dystrophy Association named Mattie its National Goodwill Ambassador for 2002. You would think that Mattie has enough to deal with, but he also lost some new friends—firefighters he had met at a celebrity softball game—in the collapse of the World Trade Center on September 11, 2001. The tragedy strengthened his resolve for peace. Somehow Stepanek acquired early in life the kind of optimism that many adults several times his age have trouble mastering. Revealing ageless wisdom, he says, "We all have life storms, and when we get through them or recover from them we should celebrate that we got through it instead of just mourning and waiting for the next one to come along and wipe us out again."[6,7]

IS PESSIMISM ALWAYS BAD? FINDING A PLACE FOR THE NEGATIVE

Just because optimism seems to be generally good, that does not mean that pessimism is always bad. As I suggested earlier, they're not two sides of the same coin. What, then, might be the redeeming value of a pessimistic attitude? Can negative expectations ever result in more favorable outcomes?

Defensive Pessimism

In psychologically risky situations researchers have discovered that many people make effective use of something called "defensive pessimism." This occurs when people purposely lower their expectations for performance, even though they've experienced past successes in similar situations. For example, going on a job interview, taking a test, going on a first date, giving an important presentation, or meeting in-laws for the first time—all these are instances when our performance will be judged. Such instances are psychologically risky in that they expose us to the possibility of failure, potentially threatening our self-esteem. In some settings the prospect of failure looms so large that it becomes imperative for us to take measures to protect our self-esteem against it.

By setting expectations *unrealistically low,* the person conducts a kind of preemptive strike against the excessive loss of self-esteem in the event of failure. It's like saying, "This test is going to be hard. No way I'll do well" despite having high grades or "I'll apply, but I probably won't get the job anyway" despite having strong credentials and past success in getting similar jobs. Defensive pessimists are buying "failure insurance." They hope they will not need it but prepare for the worst just in case.

Everyone, even someone as successful as filmmaker Steven Spielberg, can demonstrate a little defensive pessimism occasionally, as you see in this quote from a *New York Times* interview:[8]

> I'm the fraidy-cat who makes a picture and immediately assumes that nobody is going to show up the first day, and it will be reviled

around the world. That's the way I've been on every single project. Every one. When it doesn't turn out that way, I'm relieved. Relief is the largest reaction I have to a film that's well received and opens well. . . . I simply feel relief.

People who are really defensive pessimists look a lot like regular dispositional pessimists or people with a pessimistic explanatory style. But they are different. Unlike other forms of pessimism, defensive pessimism does not result in giving up when things get tough. Ironically, defensive pessimism is a strategy that helps some individuals perform well. Defensive pessimists very much want to succeed, have a history of succeeding, and are willing to put forth the effort to do so. Setting low expectations reduces their performance anxiety about failure and actually boosts their motivation. According to University of Michigan researchers Drs. Julie Norem and Nancy Cantor, defensive pessimists are in essence "playing through" potential bad outcomes before they occur, thereby gaining some degree of control over anxiety. It is as if they're bracing themselves against the impact of failure by dwelling on the possibility. In contrast, dispositional and explanatory pessimists have low motivation and become inactive and even depressed.[9]

Several studies have now shown that defensive pessimism is indeed quite useful for some people under the right conditions. In one study Norem and Cantor gave optimists and defensive pessimists a difficult mental task to perform. Prior to performing the task, they were asked how they expected to do.[10] The results showed that the two groups performed comparably, but after the task the participants in each group were given false feedback on their performance: Half were told they did well, half that they did poorly. Measures of feelings of satisfaction, control, and self-rated performance were taken. As predicted, before the task optimists felt they would do quite well, whereas defensive pessimists had lower expectations. The optimists who were told they did well felt more in control than did those who were told they performed poorly. For the defensive pessimists, success or failure did not affect their feeling of control or their enjoyment of the task. In other words, they were comfortable with success but were "cushioned" against the emotional fallout of failure.

A second study by Norem and Cantor also found that defensive pessimists and optimists set different expectations for their performance prior to performing a difficult task. Defensive pessimists felt they would perform more poorly than did the optimists—yet they performed comparably well.[11] This again suggests that setting lowered expectations at least does not hurt the defensive pessimist's performance. But does it help? One way to find this out is to prevent people from using their preferred strategies; that is, make defensive pessimists act like optimists and vice versa, to see if their performance suffers as a consequence.

This is what Norem and Cantor did in another experiment involving the defensive-pessimism and optimism scores of participants with grade-point averages of more than 3.0.[12] Before performing a set of puzzles (some easy, some difficult), half of the defensive pessimists and half the optimists were told, "Do as many as you can in the time allotted." The others in both groups were given encouragement and were told that it was expected they would perform well—something to the effect of "Hmm, given how well you've done in the past, I would think that you'd be very confident about your performance. You will probably do very well on the upcoming tasks."

The researchers believed that such encouragement would be congruent with the already high expectations of the optimists and bolster their performance on the tasks. The same encouragement was expected to hurt the performance of the defensive pessimists. According to the authors, for defensive pessimists encouragement would interfere with their strategy to keep expectations low and therefore would raise their anxiety.

Norem and Cantor found that participants performed better when their preferred style was not interfered with. Optimists performed better when they received encouragement. Their performance dropped without it. For defensive pessimists the opposite was true. They performed better without encouragement, and their performance dropped when the experimenter raised performance expectations with encouragement.

In the next chapter, I explore two additional ways that our thoughts may influence our health—through our beliefs about ourselves and through the illusions we create.

Chapter 3

The Power of Beliefs
and Illusions

Where do optimism and pessimism come from? How do we get to be one way or the other in the first place? Seligman believes that optimistic and pessimistic attitudes can originate from three influential sources:[1]

- our parents or guardians, either through heredity or through the examples that they set for us
- the criticism that we receive from our parents, teachers, or coaches that may impart optimistic or pessimistic attitudes
- our direct experiences with positive and negative events, and the meaning we attach to them

Regardless of the origin of optimism and pessimism, once formed, these dispositions tend to take root in our psyche. Thereafter they become self-fulfilling prophecies—we selectively find confirmation for our optimistic and pessimistic outlooks in the events of our lives, and we ignore events that do not confirm our perspective. We turn the "seeing is believing" notion on its head—with optimism and pessimism it's more like "believing is seeing." What we believe shapes what we see. Thus, optimism and pessimism are really about beliefs. I would even say they stem in part from our *core beliefs* about the world and our place in it. I say "core" beliefs because our characteristic ways of anticipating the future and explaining the past—whether optimistically or pessimistically—tend to be consistent across time and situations. In other words, we do not swing wildly between being strongly hopeful one day and seeing the world uncongenially the next. Not that the context or situation does not matter or is irrelevant. The context determines the degree of optimism or pessimism we experience at a given time. But we tend to remain in the same position on the optimism/pessimism scale compared to other people. This consistency suggests that a more fundamental belief system about the world and ourselves may underlie optimism and pessimism.

So it is possible that our parents or our early life experiences lead to optimism or pessimism by forming our core beliefs about ourselves and the world. That is:

Early Life Experiences—▶Formation of Core Beliefs—▶ Optimism or Pessimism

As a case in point, Elizabeth and I had very different early life experiences that may have shaped our core beliefs. I was fortunate to be raised in a stable and protected environment. My parents remained married throughout my childhood and provided my brother and me with a secure and loving home life. Although we had our share of stressful events, there were no major upheavals in our family of the sort that irrevocably altered my life trajectory. The members of my parents' church often served as surrogate parents, providing yet another layer of social stability and emotional protection as I matured. But, more important, I feel that my home and church life instilled in me a core philosophy of faith—a belief that positive outcomes are forthcoming regardless of circumstances—that not only was I taught but of which I saw examples during misfortune in countless lives of church members. All these experiences produced in me the core belief that the world is generally a safe and welcoming place, and that positive outcomes, especially during adversity, are to be anticipated. Elizabeth had a different childhood. Her parents divorced when she was a toddler, and she spent her early years with her paternal grandparents. Although she adored them and they provided a stable, nurturing environment, the absence of her parents, who were transient figures in her life, had an effect. Her mother remarried when Elizabeth was in second grade and recreated a family that now contained a stepfather. Again Elizabeth was the beneficiary of love and care, as he became her principal father figure and provided for her material needs. However,

Dr. Albert Bandura of Stanford University, one of the most influential psychologists in history, perceived self-efficacy is the conviction that you can carry out a behavior necessary to produce a desired outcome. "Unless people believe they can produce desired effects, and forestall undesired ones by their actions," says Bandura, "they have little incentive to act, or to persevere in the face of difficulties."[2] Self-efficacy is more than simply knowing what it takes to succeed in an endeavor. You may know exactly what it takes to succeed as an attorney, a stockbroker, or a public speaker. You may have even acquired some training and experience in those areas. But self-efficacy is more than knowledge, training, or even desire. It is the *belief and expectation that you can actually perform the behaviors necessary to achieve your goals.*

Self-efficacy is not self-esteem, which Bandura considers analogous to self-worth, or how much we like and value ourselves. Rather, perceived self-efficacy addresses a belief in our personal ability to achieve a desired outcome in a particular endeavor. According to Bandura, "Individuals may judge themselves hopelessly inefficacious in a given activity without suffering any loss of self-esteem whatsoever, because they do not invest their self-worth in that activity. Conversely, individuals may regard themselves as highly efficacious in an activity but take no pride in performing it well."[3]

People who are high in perceived self-efficacy:

- set goals that are important to them
- believe those goals are attainable through specific behaviors
- believe they have the ability to perform the behaviors necessary to achieve those high goals

Highly self-efficacious people believe that they have control over their circumstances—and they act accordingly. Compared to people low in self-efficacy, those who are high in self-efficacy are more likely to try new things, persist in activities even in the face of setbacks, do well in school, and age successfully.[4]

A picture of Wally "Famous" Amos should appear next to any textbook definition of self-efficacy. He has made it his mission to help people believe in themselves when the chips are down. "You can always turn your life around," he told me. "Life doesn't defeat you; it's divorce and blended families were not so common then, and Elizabeth sometimes felt stigmatized. Plus, she was separated from her older siblings, especially a sister to whom she was very attached, and from her grandparents, who had become her parental figures. In her new home she was raised in a large metropolitan area without the safety net of surrogate or extended family. Her early experiences contributed to core beliefs that the world is not necessarily hospitable, that people cannot always be trusted, and that negative outcomes likely follow adversity.

When things are going well, these core beliefs are not especially consequential. You cannot tell that there is a difference between Elizabeth and me in terms of optimism. It is only under adversity, or when outcomes are ambiguous, that these core beliefs express themselves as optimistic or pessimistic expectations or explanations.

SELF-EFFICACY: FROM BELIEFS TO ACTIONS

Research on the psychology of beliefs is about not only events and situations in our future but also what we believe about ourselves. Our core beliefs about ourselves are so important they can affect our success in setting and meeting important life goals—specifically, whether we believe we have what it takes to accomplish important goals. Our actions are strongly determined by our pursuit of goals, and those goals motivate us to act. But whether we take action, and the specific action that we take, is determined by more than our goals. It is determined also by our expectations—or our beliefs—about the attainability of those goals. Presumably, we won't pursue goals that we perceive as unattainable. We won't, for example, apply for a job as a computer programmer with no programming experience, start a catering business with no interest in or aptitude for cooking, or enter the Olympic swimming trials having just recently learned to swim. But what if the goal is realistically within our grasp? What then determines whether we will pursue the goal?

Psychologists have learned that this, our *perceived self-efficacy,* our internal sense of what we can achieve, determines what goals we set and attempt to accomplish, which activities we avoid, how hard we try, and how long we persist when faced with difficulties. According to a belief that defeats you." Amos never had self-defeating beliefs and never doubted his abilities, not even for a moment. He persisted in his dreams, confident in his skills, and walked away from those who dared try to devalue him. In the early 1960s, after a stint in the military, with a pregnant wife and a two-year-old son at home, Amos quit a managerial position at Saks Fifth Avenue because they refused him a small pay raise. He knew he was worth it and would not allow them to cheapen him. He explained to me, "Even though I had limited skills and limited education, I didn't accept that I had to work for eighty-five dollars a week, when I felt I was worth more." He said that although the situation looked critical, he made a decision to do what needed to be done, and it became a real opportunity for him.

Amos landed a job in the mailroom of the William Morris Agency and even took a pay cut from what he was making at Saks because he knew he had the talent to launch a career as a promoter. He became a successful theatrical agent, signing Simon and Garfunkel and the Supremes. He later started his own agency and, when he tired of the entertainment business, turned his promoting skills to selling the chocolate-chip cookie he remembered eating as a child. Some of his friends discouraged his narrow focus and warned that he would never survive in the location that he chose in 1975 for the first-ever free-standing chocolate-chip-cookie store. Amos applied his promoting skills to selling the chocolate-chip cookie and created a revolution in the cookie business as a result. He developed "head shots" for the cookie, distributed cookies as "calling cards" and thank-you gifts, and used a custom-designed jingle to direct customers to the store on the corner of Sunset and Vine in Los Angeles.

Amos's confidence about his ability to promote his cookie and create a market for a product was the secret to his overwhelming success. Amos admits that he was not as skilled as a businessman, and he accepts responsibility for losing his first cookie business and even the rights to use his own name in other ventures. Dark days followed, but Amos never doubted that he would rise above the circumstances. "I don't even use the word 'adversity' because I don't see adverse experiences. I see learning experiences," he told me. Yes, he almost lost his house, but he started another cookie business, became involved in various philanthropic interests—especially Literacy Volunteers of

America—and engaged in other business ventures, including a muffin company, Uncle Wally's. In 1999, a little more than sixty years old, when many people are thinking of retirement, Amos returned to Famous Amos cookies, a subsidiary of the Keebler Company, as the Director of Cookie Fun. Along the way, the Smithsonian Institution recognized his uniqueness by adding his trademark hat and shirt to its permanent collection.[5]

Perceived self-efficacy like that of Amos is one of the strongest determinants of behavior in areas as far-reaching as academic achievement, job performance, organizational functioning, and athletic performance.[6] Research in recent years has observed the following links between perceived self-efficacy and improved health:

Lower levels of depression in children. In research conducted at Stanford University, Bandura found that children with lower levels of self-efficacy regarding their social and academic skills had higher levels of depression over time.[7] That is, the children's beliefs about their ability to perform academically and socially predicted their levels of depression two years later. The link between low self-efficacy and depression was strongest among the girls.

Greater performance of health behaviors. People higher in self-efficacy for health-related behaviors are more likely to initiate and sustain those behaviors over time. Some of the behaviors shown to be sensitive to self-efficacy include doing exercise, maintaining healthy dietary habits, smoking cessation, taking medication as prescribed, managing weight, and controlling alcohol and illicit-drug dependency.[8]

Reduced biological responses to stress. In laboratory studies, research participants performing highly stressful tasks show impairments of their cardiovascular, immune, and neuroendocrine systems.[9] However, with higher levels of self-efficacy, performance on such tasks showed minimal disruptions in these systems.[10]

Improved memory among the elderly. There is great variation among the elderly in the decline of cognitive functioning (e.g., in memory). Some individuals show dramatic declines and others minimal decline or none at all. In a two-and-a-half-year investigation as part of the MacArthur Study of Successful Aging, researchers measured self-efficacy among elderly persons for performing instrumental activities (e.g., arranging transportation, being productive, ensuring safety) and for interpersonal endeavors, such as dealing with family, friends, or spouses. Men with stronger beliefs about their self-efficacy for instrumental activities were better able to maintain a high level of memory over time compared to those with lower self-efficacy.[11]

Improved management of chronic diseases. Research has established that cognitive-behavioral techniques can be effective tools for treating chronic diseases such as arthritis and diabetes. Cognitive-behavioral treatments consist of providing patients with instruction in self-relaxation, cognitive pain management, goal setting, and self-incentives. Findings indicate that the effectiveness of these techniques derives largely from their ability to enhance self-efficacy.[12]

Better control of anxiety and phobias. Dozens of studies over the last twenty years have shown that behavioral interventions that raise self-efficacy, through either active mastery training or modeling, can reduce anxiety and phobias.[13] In some studies even providing phobic or anxious patients with the *illusion* that they could control their symptoms resulted in real relief of symptoms.[14]

More effective pain control. Research indicates that a number of psychological and behavioral interventions can be used to control pain, including muscle relaxation, positive imagery, cognitive distraction, and reinterpreting bodily sensations. Pain sufferers who are taught these techniques have various degrees of success in reducing pain. A prime predictor of pain reduction is the patient's belief in his or her ability to effectively use the techniques. The patients who believed that they could successfully use newly learned pain-coping strategies indeed experienced less pain over time.[15]

Sources of Self-efficacy

Is it possible to improve your self-efficacy to meet your goals? Absolutely. Bandura has said that our self-efficacy beliefs in any particular endeavor come from at least one of four sources.[16]

Mastery experiences. Nothing builds self-efficacy like the experience of attempting a task and succeeding, even if on a small scale. Mastery experiences are just that—experiences that allow you to practice or rehearse desired skills and develop your abilities slowly. When developing self-efficacy through mastery experiences, it is useful to practice in a safe, nonthreatening environment. Once a sense of self-efficacy—the belief that you can achieve a task—is established, perseverance in the face of setbacks is more likely. Research has demonstrated that mastery experiences are by far the best approaches for enhancing self-efficacy.

Vicarious experiences (modeling). Although direct mastery experiences are best, self-efficacy can also be built by watching others perform tasks successfully. This type of behavioral modeling, as it has been called, works especially well if the "models" (those being watched) are similar to you (e.g., in ability, experience, age, gender). The idea is that if persons similar to yourself can succeed at a task, so can you, and your self-efficacy is increased.

Verbal persuasion. Sometimes our perceived self-efficacy can be enhanced when others tell us that they have faith in our ability to succeed, especially when difficulties challenge our confidence. Of course, the source of such verbal persuasion must be credible and their encouragement believable.

Physiological and emotional states. Sometimes self-efficacy stems partly from our bodies and moods. When we have to perform in threatening situations, we often experience higher levels of arousal, both physiologically (e.g., a racing heart) and emotionally (e.g., fear, sadness). High physiological arousal and negative emotions may signal to us, at least under certain conditions, that we are vulnerable to failure. In a sense, our bodies and emotions may be telling us we do not have what it takes to succeed, and our self-efficacy goes down. On the other hand, we could reinterpret the feelings associated with high arousal in ways that enhance, or at least do not diminish, self-efficacy. For example, high physiological arousal can be interpreted as augmented alertness, and anxiety can be looked at as heightened anticipation. These kinds of interpretations would tend to facilitate actions.

OPTIMISM'S OUTER REACHES: CREATING POSITIVE ILLUSIONS

When renowned basketball player Earvin "Magic" Johnson announced in 1990 that he had HIV, he was probably the only person on earth who thought he would live another year. Having HIV, the virus that causes AIDS, was then considered a death sentence. Everybody knew that. Everybody but Magic Johnson. Doctors told him that his attitude would be key in beating his disease. Magic says, "And you know me, when the doctor said that, I said, 'Oh, shoot. It's beat then.'"[17]

Most of us have some unrealistically positive views of ourselves relative to other people. We think that our chances of being happy, of having a successful career, a good marriage, or talented children, will be better than the chances of our peers. We also feel that the chances of having bad things happen to us are considerably fewer. We hold these positive beliefs about ourselves even though, statistically speaking, they are not likely. Thus, these beliefs are a type of positive bias regarding our lives.[18]

Because of the consensus that the hallmark of mental health and emotional stability is the accurate construal of reality, behavioral scientists have debated for years whether such positive biases are emotionally good or bad. Some research indicates that believing we have more control over our circumstances than we actually have may be emotionally beneficial. In several laboratory experiments people who are depressed have been shown to have more accurate perceptions of their degree of control than nondepressed people have. For example, in experiments on learned helplessness, when participants performed tasks over which they had little control, depressed people accurately assessed themselves as having very little control. Nondepressed people, on the other hand, believed they had control even in circumstances where they actually had none at all.[19]

Other research indicates that even the perception of social skills differs between depressed and nondepressed people. Specifically, depressed people are often very accurate judges of their social skills. Their self-assessments more closely mirror the perception that others have of them. Nondepressed people, however, dramatically overestimate their skills, believing themselves to be more likable, persuasive, and appealing than others think they are.[20] Depressed people often view themselves accurately, but they may pay an emotional price for this insight.

The work of Dr. Shelley Taylor, a psychology professor at the University of California at Los Angeles, has also challenged the notion that perceptual accuracy is always good. She believes that, particularly during times of extreme trauma, distorting reality a little can be adaptive and helpful.[21, 22] Taylor arrived at this counterintuitive theory in the early 1980s while interviewing breast-cancer patients for a study. She and her colleagues were interested in identifying the factors that helped the women return to their previous levels of functioning after the traumatic and life-threatening experience of cancer.[23] The research examined, among other things, the idea that women who had enhanced feelings of control over their cancer would show better emotional and psychological adjustment to the disease. This hypothesis grew out of the research showing that controllable stressors produce fewer negative emotional effects than do uncontrollable ones. Taylor reasoned that the *belief* in control could have the same beneficial effects as actually *having* behavioral control.

One of her many findings was truly startling and on the surface appeared to contradict a long-held view of mental health: Many of the women who were coping best with their cancer had extremely and unrealistically positive views of their control. Taylor writes, "Many women expressed the belief that they could personally control the cancer and keep it from coming back. Others insisted they had been cured, although their records showed them to have progressing illness. Despite the fact that these beliefs were inconsistent with objective medical evidence, they were associated with [good] mental health [in these women] and not psychological distress."[24]

These findings, and those of other researchers, led Taylor to propose the idea that "positive illusions"—beliefs that represent mild distortions of reality—could actually provide some mental-health protection against traumatic experiences and could ultimately affect our physical health as well. One of the key features of positive illusions is what is known as "unrealistic optimism."

Earlier I talked about Art Berg and his incredible optimism even while lying on the highway waiting for an ambulance after his car accident. He told me that during his recovery he always held out hope that things would change. He said that although he realized that some of his hope was not based in reality, it kept him from being depressed and despondent. He said he felt down, but never hopeless. The reaction of his doctors to this hopeful disposition is interesting: They were frustrated by what they perceived to be incessant optimism, and their response was to isolate him. They put him in a room by himself and restricted his family visits. At the time he wasn't sure why he was being treated this way, but later he happened to gain access to his medical charts and was surprised to find what a doctor had written. The doctor felt that Berg's problem was that he was experiencing "excessive happiness" that was damaging his progress. The doctor wrote that he felt that Berg was in a state of denial and needed to accept his condition. Berg explained to me, "I was just not willing, in their words, to accept it. That didn't mean I had an unrealistic view of where I was. To accept it was to give in to the sense of hopelessness. To me that was kind of the death knell."

Taylor and her UCLA colleague Dr. Margaret Kemeny investigated the health effects of positive illusions in people facing the threat of death from HIV. In several remarkable studies involving some of the 1,400 participants of the Multicenter AIDS Cohort Study (MACS), a large, long-term study of the natural history of AIDS, the impact of positive illusions has become crystal clear.

The first study involved 238 men who were HIV positive and 312 men who were HIV negative.[25] The men were asked questions about their dispositional optimism and about their optimism specifically related to AIDS (AIDS-specific optimism). Concerning the latter, they were asked how optimistic they were about their chances of staying free of AIDS. The most dramatic finding in the study was the presence of strong positive illusions and unrealistic optimism among the men who were HIV positive. Amazingly, these men, who had already contracted HIV, *were more optimistic about not developing AIDS* than were men who were HIV negative. The HIV-positive men were *unrealistically* optimistic, since in fact their chances of developing AIDS were actually *greater,* not lower, than those of the men who were HIV negative. This finding supports the idea that unrealistic optimism arises in response to extremely threatening life conditions, perhaps as a way of coping with distress.

There were other findings as well. Participants showing both AIDS-specific and dispositional optimism had less psychological distress than did the other men in the study. Conversely, some participants reported what was called "fatalistic vulnerability" (a kind of pessimism) about the course of AIDS (e.g., believing that AIDS could not be avoided). Individuals high in fatalistic vulnerability experienced the greatest psychological distress, worried more about AIDS, and perceived themselves as having less control over AIDS than did those with lower fatalistic vulnerability and those who were more optimistic.

An interesting question arose: Did the optimism precede the HIV infection, or did HIV lead to unrealistic optimism, as the researchers believed? It could be that the more optimistic men exhibited riskier sexual behaviors, thereby increasing their risk for HIV. As a check for this possibility, the researchers conducted an analysis of those men who were tested for HIV but chose not to learn the test results. Among these men there were no differences in AIDS-specific optimism between those who were HIV positive or negative. It appears, then, that it was the *knowledge* of HIV-positive status (the threatening event) that seemed to trigger the illusion of invulnerability. According to Taylor and Kemeny, the *illusion of control* helps people adjust and adapt emotionally to events that could otherwise be overwhelming.

Is it helpful or not to be unrealistically optimistic when faced with a terminal illness? The influential work of Dr. Elisabeth Kübler-Ross has guided thinking on this issue for more than three decades.[26] According to Kübler-Ross, a person facing a terminal illness goes through predictable stages in coming to terms with death. The final stage in the dying process is one of resignation and acceptance—indeed, a realistic acceptance—of one's inevitable mortality. Kübler-Ross believed that this final stage of acceptance was crucial for the individual to achieve psychological peace at life's end.

But many patients with terminal illnesses steadfastly refuse to accept their plight, sometimes to the very end. They believe that they, or their doctors, will be able to fight the disease and win. The question is, then, is it good to have an unrealistically optimistic attitude? Will that person be worse off emotionally if and when the realities of a serious illness dash all hopes?

Magic Johnson shocked fans and foes alike when he retired from basketball on November 7, 1991. Almost every sports enthusiast can remember when he or she heard the announcement that basketball was losing Magic because he had contracted HIV. Johnson wanted to spend the time he had left with his family and educate others about AIDS. It is an understatement to say that this was a sad time for Johnson, who for the first time was without that ever-present irresistible smile.

But in a short time Johnson began making more public appearances and speaking out about AIDS. His mood was once again upbeat and positive, like the old Magic. And, amazingly, despite the lack of a cure for AIDS, Magic honestly believed he had a good chance of beating it. In his 1992 autobiography, *My Life*, he writes about his attitude concerning HIV:

> One thing that really makes me mad is when people say that I'm in denial about having HIV. Maybe they think that way because I continue to be upbeat and optimistic. But I'm not going to crawl into a hole. The truth is, I don't have bad days. I don't wake up in the morning and think I'm going to get AIDS. . . . There have been moments of sadness about the virus, but not many. I've always been that way, thinking positive, with a bright outlook on life. . . . To me,

HIV is just another challenge. I have the same attitude I always had on the court. I'm going all out to fight it. I haven't lost too many battles in my life. . . .

Was it good for Magic to be so optimistic about his prognosis? Is it better for people such as Magic, with serious illnesses, to be more accepting of their conditions?

Taylor and Kemeny are also providing some surprising answers to these questions. Two intriguing studies, conducted in collaboration with their colleague Geoffrey Reed, have suggested that, in the context of terminal illness, unrealistic optimism may prolong life, while realistic acceptance may shorten it. The first study involved 74 gay men with AIDS from the MAC study, who were diagnosed an average of twelve months before entering the study.[27] Biological and psychological assessments were made, and the men were studied for more than two years. One key measure was called "realistic acceptance," a kind of negative (though realistic) expectation about their future with AIDS. People with high realistic acceptance endorse statements such as "I try to accept what might happen" or "I prepare myself for the worst."

From the perspective of Kübler-Ross and others, realistic acceptance is exactly where AIDS patients should be at certain stages of their illness.[28] This is especially true since the study was conducted before the widespread use of combination drug therapies. Persons scoring high on realistic acceptance were "essentially acknowledging the likelihood of their risk for death, whereas those who score low are not engaged psychologically or behaviorally with the final stage of life." The UCLA study was designed to test the value of realistic acceptance. The findings were stunning. *Men who realistically accepted their condition—that is, had negative expectations about it—lived an average of nine months fewer* than those who did not have negative expectations. Figure 2 shows the survival time of the men who were low and high in realistic acceptance.

Just ten months into the study, differences between the groups were already emerging. Those who realistically accepted their fate were already showing a higher death rate. The survival differences between the groups were not due to poorer health behaviors or a failure to seek medical attention by those who had accepted their plight. The differences were also not due to less social support (see Chapter 6) or

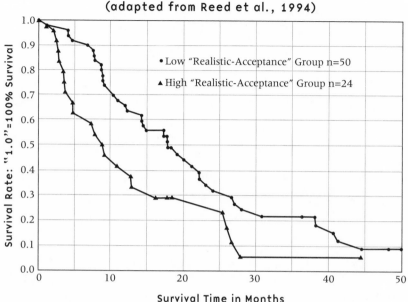

Figure 2: Acceptance and Survival Time in AIDS
(adapted from Reed et al., 1994)

• Low "Realistic-Acceptance" Group n=50

▲ High "Realistic-Acceptance" Group n=24

Survival Rate: "1.0"=100% Survival

Survival Time in Months

more negative moods in the realistic-acceptance group. Instead it appeared to be something about the differences in the expectations of the two groups that set them apart.

In the second study by Reed, Kemeny, and Taylor, 72 HIV-positive gay men were studied over the course of two and a half to three and a half years. About 35 of the men had experienced bereavement within the year prior to the study, losing either a partner or a close friend. Assessments of expectations about AIDS were obtained at the beginning of the study, when all of the men were free of AIDS-related symptoms. However, over the course of the study, the illness progressed. Men developed symptoms such as persistent diarrhea, unintentional weight loss, persistent fevers, and persistent night sweats. Men who had reported high negative expectations about the course of AIDS (e.g., believing their disease would progress) developed symptoms at a significantly higher rate than did those with low negative expectations. This was especially true for men who had experienced bereavement. Symptoms of AIDS developed in approximately 66 percent of the bereaved men who also had negative expectations, while only 40 to 50 percent of the men in the other groups did so. The results remained unchanged after taking into account immune-system status, use of the drug AZT, substance abuse, depression, and high-risk sexual behavior.[29]

It's hard to argue with success. Magic Johnson is often quoted as saying, "I thought I'd be here." He is the beneficiary of a complement of new HIV/AIDS medications, a better understanding of the course of the disease, the compassion of family, abundant financial resources, and the indomitable, magical Magic Johnson worldview. He says of his diagnosis, "I'm a guy of challenges and that was another challenge in my life, to see if I could conquer it, and so far so good."[30]

CONCLUSION TO PART I

Many religions and philosophies emphasize turning one's thoughts away from the future, consciously being "in the moment." This practice can take many forms, but the goal is always the same: to attempt, whenever possible, to be mindful of one's present experience, to be fully absorbed in the activity at hand—not to ruminate unnecessarily about the past or worry about the future. In fact, mindfulness meditation, which teaches a focus on the present, is a powerful technique for relaxation, stress management, and the treatment of some chronic diseases.[31] Staying in the moment makes perfectly good sense, given that the only time that really exists is now. The curtain on the past has been closed, and the future awaits unveiling. Unfortunately, most of us rarely attain present-moment focus. Our minds transport us from the here and now to the what has been, the what will be, the what has gone by, and the what is yet to come. Given our propensity to spend a good deal of our lives interpreting the past and anticipating the future, it would be wonderful if we could do it in a way that does not leave us downhearted, defeated, or filled with dread.

Our worldviews are shaped by the conditions of our early lives and our current circumstances. But, as is illustrated in the last three chapters, we are not at the complete mercy of our backgrounds and circumstances, and we can exert some control over how we respond to events. By exerting this control and making conscious decisions about our expectations and explanations, we become active participants in shaping our health.

There is much that we still need to learn about optimistic expectations and explanations. In particular, although optimists are on average healthier than pessimists—that is, they have fewer illnesses and better biological functioning—we still need more research demonstrating that if people are taught to have a more optimistic outlook, they will live longer. Those types of studies are under way at universities around the country. Yet we know enough now for you to take action, to reevaluate your expectations and explanations about events. Will you be healthier and live longer as a result? Quite possibly. Will you have a higher quality of life and more emotional well-being? Most certainly. Some people like to say, "I'm not an optimist or a pessimist; I'm a realist." But it's not clear what a realist is when most events in our past can be interpreted a variety of ways and the future is a complete tabula rasa. It's all a matter of perspective. So why not choose the perspective that, although firmly grounded in the facts, is most uplifting, most fulfilling? Regardless of its effects on longevity, viewing events in a more optimistic light may just be a better way to live.

Optimism, self-sufficiency, and positive illusions are only some of the psychological and behavioral elements (i.e., thoughts and actions) of the new definition of health. The next two chapters illustrate that what we do after we're exposed to major adversities or trauma can determine the long-term health consequences. The health effects of revealing or concealing traumatic experiences are the focus.

PART II

THOUGHTS AND ACTIONS 2: CONCEALING AND REVEALING TRAUMA

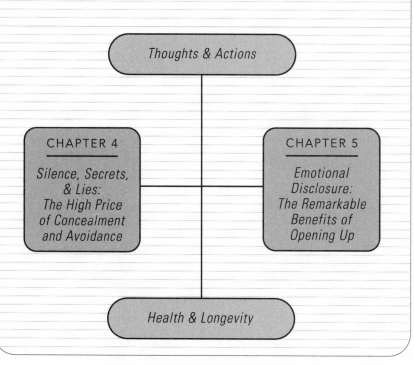

Thoughts & Actions

CHAPTER 4

Silence, Secrets, & Lies: The High Price of Concealment and Avoidance

CHAPTER 5

Emotional Disclosure: The Remarkable Benefits of Opening Up

Health & Longevity

For secrets are edged tools . . .

—John Dryden

Silence augmenteth grief . . .

—Fulke Greville

T he threat of traumatic experiences is ever present in our society. From physical and sexual abuse, the untimely and shocking deaths of loved ones, major accidents, serious illnesses, and
random violence in our neighborhoods to something as commonplace
as divorce—we all bear some degree of risk. Behavioral research has
discovered that recovery from trauma may be partially based on how
inclined victims are to reveal their trauma. Disclosing one's deepest
thoughts and feelings about traumatic experiences can serve as a behavioral antigen, helping to prevent the negative health consequences
of trauma and countering the pathogen of concealment.

The idea that it can be helpful to talk about problems and situations that bother us is far from new. Much of psychotherapy is built
around talking. Techniques such as cognitive behavioral therapy, for
example, involve the psychotherapist and the patient discussing how
a situation can lead to healing. Expressions like "confession is good for
the soul," "getting something off my chest," and "blowing off some
steam" exemplify the belief that verbal communication of emotionally
charged experiences can have salutary effects. But recent research
now suggests that these common notions of emotional catharsis are
only partly correct—and in some cases outright wrong. This research
has taken us beyond folk wisdom, illuminating the complete consequences of concealing or revealing traumatic experiences. These consequences, which are simultaneously different from what we've
always thought and more powerful than we've ever imagined, have
striking health implications.

A Breath of Air

An October field trip with your daughter's kindergarten class to a pumpkin farm is not supposed to turn into disaster, but it almost did for our friend Angie. Uncharacteristically, Angie headed out for her chaperoning duties without thinking about the hay, ragweed, dust, and dirt she would encounter that day. Any one of those can provoke an asthma attack in Angie; together they can be deadly.

Angie's chest began to tighten, and she started to have difficulty breathing. No inhaler. No cell phone. The consummate trouper, she resisted disrupting the children's fun by asking for help. Angie has had asthma since childhood and had been in precarious situations before. This time, however, she recognized that she was crossing a dangerous line. Another mother on the trip, who happened to be a physician, got Angie back indoors to a little country store, where she drank warm cider and coffee and began to feel better. This minor drama could have had a multitude of endings, not all of them good.

Another friend, Shanti, begins every day by confronting her asthma. After she awakes each morning, it takes her fifteen or twenty minutes to awaken her lungs. Until she has cleared them, she cannot take a complete breath. Each attempt brings on congested coughs and wheezing. This daily morning ritual restricts her lifestyle, including her intimate relationships. Sometimes even laughing can initiate a coughing spasm. Shanti must also limit her outside activities during the winter, because cold air is a trigger for her asthma attacks. Although she lives in a hot, humid part of the country, air-conditioning is out of the question.

Imagine what it must feel like to fight for a breath. Most of us don't even think about breathing; we just do it. What if making sure you could breathe were something you had to consider every day? Holding your breath for a few seconds can't begin to approximate the terror that people with asthma must feel during an attack. You can almost appreciate the feeling as you watch an actor swim to the surface from underwater depths. You see his cheeks puffed and lips pouted as he tries desperately to propel himself through the water. As you watch, your own breathing becomes short, and you share his first deep and grateful gasp of air as he breaks the surface of the water.

What if you never got that feeling of relief? Fifteen million people in the United States face that prospect daily. More than five thousand die from asthma each year. The incidence of asthma has been increasing for decades, especially in children.

Dr. Joshua Smyth's unique investigation was designed to determine whether simply writing about stressful experiences could reduce the symptoms of a chronic disease like asthma. Several studies had shown the power of structured writing to prevent illnesses and doctor visits for healthy people, but Smyth and his colleague Dr. Arthur Stone of the State University of New York at Stony Brook set out to put writing to a more rigorous challenge. They designed and conducted a study, published by the prestigious *Journal of the American Medical Association*,[1] that may open the door to new nondrug treatments for chronic illnesses.

The Writing Cure: Smyth's Groundbreaking Study

Before I talk about the Smyth study, let me provide a little context. Describing his study for you here, at the beginning of this part of the book, is a bit like showing you a movie that first reveals the ending, then supplies the supporting details. You see, the Smyth study is where science is now; it is the most recent and significant achievement in a field that's been on a voyage of discovery for some time. This part of the book is about that voyage, one that has taught scientists two aspects of traumatic experiences: (1) traumas that are concealed and not disclosed can have major and long-lasting emotional and biological consequences, and (2) writing about such traumas can be healing.

A well-trained young researcher, Smyth acknowledged to me that he was initially somewhat skeptical about the reported effects of disclosive writing, but after he scrutinized the published research, he became intrigued. One issue that kept nagging him was the fact that most of the research showing the positive effects of writing about stressful events had been done on healthy people. Tests had never been done on people suffering from serious chronic illnesses. Smyth wondered if writing could actually help people who were already ill and if improvements in health could be objectively determined. To put writing to this more rigorous test, he selected two chronic illnesses:

asthma and arthritis. Smyth explained to me that he chose two diseases rather than one because "if we found [positive results] in just one disease, we would be open to the potential criticism that it was something unique about those kinds of folks or that illness. We wanted to see how strong the effects would be by testing [writing] in more than one disease population." As a psychologist doing research with medical patients, finding physician collaborators who were experts in asthma and arthritis, and who were also interested in psychological and behavioral options for treatment was important for Smyth. He found two who were interested in his hypothesis and willing to take part in the rigors of conducting an investigation, one a rheumatologist, the other a pulmonary specialist.

Smyth's research staff placed advertisements in local newspapers and posted flyers at local hospitals and doctors' offices, inviting people to "participate in a study of your daily experience of illness." They enrolled 112 participants out of 465 respondents after telephone screenings to exclude people with certain psychiatric disorders, medication regimens, or time constraints that might jeopardize the study. They explained the requirement to write for twenty minutes without stopping on three consecutive days.[2] Once enrolled, participants were randomly assigned to one of two groups: One group was asked to write about "the most stressful experience they had ever undergone," and the other group was asked to write about their daily activities in a kind of time-management exercise. To keep people from predicting a certain outcome and thereby influencing the results, all participants were told that the study was simply about their experiences with stress.

Smyth told me that some of his participants, and many researchers, view the writing intervention as trivial, since it's just twenty minutes a day, three days a week. But he has found that outside of the twenty-minute sessions, and maybe even after the experiment is over, participants are still thinking about and processing whatever it is they had been writing about. Smyth says that the writing experience "is an ongoing process, not one that is nicely packaged into three twenty-minute sessions."

In Smyth's study, disease symptoms were evaluated at intervals of two weeks, two months, and four months after the writing exercise and compared to evaluations done a week before the study began.

The results were striking. Among the asthma patients, those who wrote about stressful experiences showed significant improvements in lung function (i.e., they could breathe more easily) at the four-month follow-up, whereas those in the control group showed no improvement.[3] Similarly, although arthritis patients in the control group showed no improvement in symptoms, those in the experimental group had a 28 percent improvement. Overall, of the 70 asthma and arthritis patients in the experimental groups, 47 percent showed clinically significant improvements, whereas only 24 percent of the 37 control patients evidenced such improvement. Why was writing about stressful experiences so helpful for these patients? The answer to this question is complex, but, in its simplest form, one can draw an analogy between concealing and revealing stressful experiences and the concepts of pathogens and antigens.

In medical science pathogens are agents, such as viruses and microbes, that cause disease. Antigens, in contrast, assist the body in fighting such foreign agents. Phenomena also exist that can be considered *behavioral* pathogens and antigens, and these, too, have health implications. The concept of behavioral pathogens was perhaps first used by Dr. Joseph Matarazzo to describe behavioral risk factors for diseases such as smoking, physical inactivity, and excessive alcohol use.[4] The research described above and later in this part of the book points to yet another behavioral pathogen—the concealment of traumatic experiences. It taxes us mentally and physically, making us vulnerable to disease. The remarkable findings from Smyth and others suggest that the act of revealing these experiences can be health enhancing—serving as a kind of behavioral antigen. Before I describe the benefits of opening up, let's look at what happens when thoughts, secrets, emotions, and traumas remain undisclosed.

Chapter 4

Silence, Secrets, and Lies: The High Price of Concealment and Avoidance

Linda Ellerbee is an award-winning, highly respected journalist, columnist, author, and television producer. Included in her impressive resume is a twelve-year stint as a network correspondent and news anchor at NBC, covering Congress, the White House, presidential campaigns, and national political conventions. She is a popular public speaker and president of Lucky Duck Productions, which produces outstanding programs for network, cable, and public television, including the unique news programming for children on Nickelodeon, *Nick News*. She successfully created several innovative television specials on topics such as AIDS, breast cancer, and abortion. Involvement in so many activities takes tremendous energy, especially if you do them as well as Ellerbee. But breast cancer stopped her in her tracks for a while. About her February 1992 diagnosis she explains, "In the beginning, right after the diagnosis, I could think of nothing else. I would try to write on my computer, and all of a sudden it would seem as if a giant message popped up on the screen saying YOU HAVE CANCER."[1]

Some traumatic experiences are so aversive and unsettling that the last thing we want to do is think about them. Yet, as Ellerbee recounted, thoughts and images about the experience—what psychologists call "intrusive thoughts"—come crashing in out of the blue. As unsettling as these thoughts are, many scientists believe they are actually a part of the brain's effort to move us toward psychological healing.[2] Of course, our initial reaction is to turn off intrusive thoughts and avoid them altogether if we can. In fact, in some forms of psychotherapy, the central goal is to eliminate negative thoughts and feelings. But research suggests that the manner in which we attempt to alter our thoughts and emotions after a trauma can determine whether we succeed in doing so. Under certain conditions our attempts to control thoughts, feelings, and words can backfire. We can unwittingly enhance the very things we're trying to get rid of. And, unfortunately, when we try to conceal the experience of past traumas, ill health can be a consequence.

Don't Even Think About It

I would like you to try a little experiment. For the next thirty seconds, I want you to put this book down. Sit quietly and think of anything you want—that is, anything but "white bears" or related topics. Avoid thinking about bears, snow, the Alaskan tundra, or anything else in the white-bear universe. Okay. Book down.

How did you do? My guess is that you probably thought more about white bears in those thirty seconds than you've thought about them all day, all week, or maybe even all year. Some of you had visual images of white bears, others may have simply seen or heard the words "white bears" in your thoughts, or others may have seen related images.

Isn't it ironic that the very thing I asked you not to think about suddenly began to play a large role in your consciousness? Yet your experience is exactly what psychologist Dr. Daniel Wegner, author of *White Bears and Other Unwanted Things,* would have predicted. The little experiment I asked you to do is based on Wegner's work and theory—which is appropriately called "ironic process theory." It explains that when we attempt to suppress or control our thoughts or emotions, we unwittingly set into motion mental processes that eventually prevent that very control. According to Wegner,

> It seems perfectly sensible, for instance, that a person trying to abstain from alcohol might begin by trying not to think about drinking. And, too, it stands to reason that a person who feels overanxious might try mentally to relax, or that a depressed person might hope to

remedy the problem by avoiding sad thoughts . . . But the simple decision to try to control our minds can sometimes lead us wildly out of control—turning what we thought was an antidote for our mental malaise into the very poison that creates it.[3]

This does not mean that we have no control over our thoughts. Of course we do. However, under certain conditions, attempts at mental control can produce what we least want to experience. According to Wegner, this happens because the mind uses two complementary but oppositional processes to help control thoughts: the *intentional operating process* (what I will call the "operator") and the *ironic monitoring process* (what I will call the "monitor"). The operator works on the conscious level, helping us think about what we *want* to think about. The monitor, on the other hand, works unconsciously, telling us when we're having or are about to have thoughts that we do not want, so we can immediately correct them.

As mentioned, the operator and monitor are complementary. If, for example, you're driving and looking for a street name, the operator keeps that name in the forefront of your mind while the monitor is constantly registering and rejecting the wrong street names. If you're trying to find the name of a store in the Yellow Pages, the operator is actively looking for the name while the monitor is telling you that all the other names you see are not what you're looking for, urging you to keep moving and not waste time.

The monitor and operator work together similarly when we want to avoid thinking of something. To return to the example of the white bears, when you're trying to avoid thoughts of them, the operator helps you purposely think about other things—cars, food, work, movies, or anything *but* white bears. The operator is trying to give you what you want, in this case distracters. At the same time the monitor is working unconsciously, looking for anything that might remotely resemble white bears, so that it can tell you to skip over those things. And it is this monitoring process that makes it nearly impossible to completely avoid certain thoughts when you're trying to. Here is how the operator and monitor might be working when you're trying to avoid thoughts of white bears:

OPERATOR, trying to distract you: I wonder if anything good is coming on television tonight?

MONITOR: Be careful, sometimes the science station has shows on WHITE BEARS, and you don't want to think about them.

OPERATOR, trying again: Okay, better think about food. What should I have for dinner?

MONITOR: Watch out, you love fish, and so do WHITE BEARS.

OPERATOR, trying one more time: Just think about hanging out at the beach, watching the waves.

MONITOR: This is good. Certainly no WHITE BEARS at the beach.

Of course, the operator and monitor don't always use words; sometimes they just use images, as when the operator thinks of dinner and the monitor produces an image of a white bear with a fish in its mouth. The point is that the interplay between the operator and the monitor makes it hard for you to completely banish thoughts about the white bears when you try to, because the monitor occasionally intrudes just to make sure you're not thinking about them—which of course *causes* you to think about them, if only briefly.

Most of the time this process serves us well when we're trying to avoid certain thoughts or feelings. The occasional intrusions of unwanted thoughts into our consciousness keep our operator on its toes. They remind us that we need to distract ourselves more effectively. However, Wegner and others have found that when people are under stress or are mentally challenged, the monitor can overwhelm the operator, flooding the consciousness with the unwanted thoughts or moods, producing the following effects: Thoughts we try to suppress are more likely to occur. Emotions we try not to experience are more likely to be felt. Physiological changes indicative of anxiety, such as perspiration, are experienced when we're trying to relax.[4]

Once we stop suppressing unwanted thoughts, we may experience a rebound effect, whereby we think *even more* about the suppressed topics than we ordinarily would. Chronic thought suppression might in fact be a risk factor for depression and anxiety.[5] Unfortunately for depressed individuals, attempts to suppress negative thoughts often backfire, leading to an increase in the negative thoughts they're trying to suppress.[6]

The Whole Truth and Nothing but the Truth

A difficult time occurred for me at NIH just before I formally announced that I was leaving for a position at Harvard University. I had completed the confidential negotiations with Harvard about two weeks before, but I needed to delay the announcement. Those were among the hardest two weeks of my time in Bethesda. Essentially, I was keeping a very big secret. What made it particularly difficult is that I was keeping it not only from the behavioral-science community but also from the people with whom I interacted daily at work. Meetings with staff stimulated a flood of thoughts about my departure, especially given that we were frequently planning activities for months or even a year in the future, when I knew I would not be at NIH.

Wegner and others have discovered that keeping secrets is more difficult than we ever imagined. We must constantly monitor what we do and what we say, especially in the presence of the people from whom we're keeping the secret. In addition to forcing us to watch our behavior, secrecy requires that we suppress thoughts about the secret, lest we commit a slip of the tongue or provide hints of our deception through body language. My operator and monitor were really having a go at each other. I did not want to think about leaving for Harvard, and my operator was giving me all kinds of other things to think about. But my monitor was constantly letting me know when situations or thoughts were relevant to my leaving so I wouldn't slip up, as when planning future activities with staff, or running into co-NIHers in the hall—of course each encounter stimulated thoughts about leaving. For two weeks, every time I ran into people I knew, my monitor yelled out, "You're leaving NIH! Don't tell them yet!" Meetings were especially bad, since every time I looked someone in the eye, my monitor went off.

All this secret keeping involves work and effort, and over time it can take a biological toll. Incredible findings from the University of California at Los Angeles illustrate how powerful concealment can be. Using data from the Multicenter AIDS Cohort Study, researchers explored whether HIV infection would progress more rapidly among HIV-positive, but otherwise healthy, gay men who concealed their homosexual identity than among those who did not. After nine years it was found that the more the men concealed their sexual identity, the faster their HIV progressed. Amazingly, there was a dose-response relationship between degree of concealment and immune status, the likelihood of being diagnosed with AIDS, and AIDS-related mortality. The greater the concealment, the higher the rates of disease and death.[7]

The gay men who rated themselves as "closeted" were essentially keeping a very big secret about an issue that was an important part of their lives. Such concealment clearly had tragic effects on their bodies. But what about people who, instead of trying to conceal something, simply invent a substitute story—what is commonly referred to as lying. Are they in even more trouble? On the face of it, it might appear that maintaining a running fabrication would be harder than carrying a secret, what with keeping the facts straight and avoiding contradicting oneself. Seems like more, rather than less, work than keeping a secret. But psychologically (not morally) speaking, just the opposite is true. With lying, Wegner believes that the fabricated story serves as a substitute or distracter for the concealed truth.[8] A fabricated story allows its creator to construct a reality that takes the mind away from the secret and into a world where the secret, on a public and partly psychological level, does not exist. It's the difference between the veritable, truth-concealing liar and the tormented, yet truth-desiring concealer. Because lying has components of secret keeping, it does exact a toll and it does lead to many *other* problems. It's just that the emotional consequences may be less.

So the liar relieves some of the strain of keeping a secret by creating a new story that serves as a distraction. A secret-keeper, however, has no distracters—nothing to mentally fill in for the secret; therefore, the secret-keeper is "left with nothing to think about but the secret itself."[9] The secret-keeper then works to suppress thoughts of the secret, especially in the presence of the relevant audience. Of course, this attempted suppression only causes intrusive thoughts about the secret to manifest themselves in the psyche of the secret-keeper.

My own secret about leaving NIH caused intrusive thoughts to run rampant in my mind. I could not stop thinking about it, especially at work. I didn't lie about it, and fortunately I was not pressed into a corner where I had to decide between revealing my secret and fabricating a story. Yet the decision to withhold my announcement was extraordinarily uncomfortable. Every time I would have a face-to-face or e-mail conversation about the future of some activity I would be involved in at NIH, I would feel either extremely guilty for not revealing it then and there or equally fearful that I just might.

But my minor thought intrusions pale in comparison to some of the devastating traumas people have experienced and the work they've done to keep the secrets to themselves. Concealing traumatic experiences is actually common. Most rape victims, for example, keep the fact of having endured this crime to themselves, neither talking to others about it nor reporting it to the authorities.[10] Some traumas are more likely to be discussed than others, perhaps because some are more socially acceptable than others. For example, experiencing the early death of a family member is discussed more than are childhood experiences of sexual trauma, parental divorce, or violence.[11] As if the trauma itself were not enough, concealing it can actually create additional problems for the victims. More on that shortly.

What, Me Anxious?

One of the things that most impresses me about behavioral and social research is the complexity of the questions that are addressed and the level of methodological creativity necessary to answer them. There is no area where this is truer than in research on emotions. We all experience emotions, but to measure them precisely is challenging. Unlike biologists peering into a microscope, behavioral scientists have to rely on people's "self-reports"—what they say they're experiencing emotionally. There is really no other way to assess what someone is feeling. These self-reports of emotion can be very reliable and, as will be described in the chapter on emotions, are often strongly predictive of a host of health outcomes. But self-reports can sometimes be highly unreliable or even false. Sometimes people want to disclose their true emotions, and sometimes they want to keep them hidden. In the research laboratory, being able to tell the difference can determine whether scientific results are valid or not. This is where scientific creativity and methodological rigor must come to the fore.

The complexity of research on the disclosure and concealment of emotions is vividly illustrated in studies of anxiety, one of the most toxic emotions. People who report high levels of anxiety are expected to show behavioral and physiological responses that corroborate their self-reported feelings. People who say they are low in anxiety should perform well on tasks with which anxiety might interfere, and they should show physiological responses free from a stress reaction. However, this is not always the case. Studies have found that some people who said they were not anxious exhibited physiological changes in response to stress that suggested they actually were highly anxious. There was a mismatch between what they reported and what their bodies were showing. This type of counterintuitive finding was once a big problem for research on emotions. What does the self-report of anxiety mean if it doesn't line up with other indicators of anxiety, like behavior and physiology? The scientific innovation that was needed to address this quandary was achieved in a psychology laboratory at Yale in the 1970s.

Psychologists Drs. Daniel Weinberger, Gary Schwartz, and Richard Davidson figured out a way of distinguishing people who were genuinely low in anxiety from those who said they were but in fact were not. To do so they administered two questionnaires. The first was a standard measure of anxiety, which classified people as high, moderate, or low in anxiety. Among those who said they were low in anxiety were some who actually were highly anxious. So a second questionnaire was given to tease out the genuine low-anxious people from the impostors. This second questionnaire measured a characteristic called "defensiveness," which differs from the customary definition (i.e., quick justification of actions in response to a perceived attack). In this context researchers label persons defensive if they present an idealized or perfect image of themselves. Highly defensive people would say, for example, that they *never* resent being asked to return a favor or that they *never* insist on having things their way. The Yale psychologists reasoned that when highly defensive people complete an anxiety questionnaire, a large percentage of them will report being low in anxiety, even if they're not, since this would be consistent with their need to project a perfect image of themselves. People who fitted this profile of saying that they had no anxiety but who were also highly defensive were labeled "repressors."

Repressors generally cope with feelings of anxiety by denying that such feelings exist and suppressing disturbing thoughts. Their avoidance of the negative is so excessive that it can affect their memory of negative events or feelings: When presented with lists of positive and negative words, repressors remember fewer of the negative words than do nonrepressors. Repressors also report fewer negative events from their childhoods than do nonrepressors.[12] Repressors make it sound as if they had rosier childhoods than nonrepressors did, even though other information does not corroborate their reported memories.[13] The repressors' pattern of emotional suppression and defensiveness gets them into physiological trouble. Compared with people who are truly low-anxious, or high-anxious, repressors generally have:

- higher heart rates and blood pressures[14]
- larger biological responses to stress[15]
- elevated cholesterol, cortisol, and glucose levels[16]
- reduced immune-system functioning[17]

Although far from conclusive and still very controversial, some data suggest that repressors may be at greater risk for cancer and that cancer, once diagnosed, may progress more rapidly in this group.[18]

Traumas Concealed

The research on repression and anxiety provided a scientific foundation for the notion that holding back or concealing emotions could have biological and health consequences. But even repression researchers likely had no idea of the magnitude of such consequences when the concealment was about a particular type of emotional experience—extreme trauma.

Although far from its being a trauma, when I was keeping the secret of my imminent departure from NIH, I began noticing changes in my mental and physical state. For the most part I just didn't feel like myself. Most of the time I'm rather easygoing and relaxed. But during this period I vacillated between feeling agitated and irritable, or exhausted and melancholy. I experienced tension headaches, stomach pains, and vague body aches, as if I were coming down with the flu. My sleep was fitful and populated with repeated dreams about telling people that I was leaving. It would be an exaggeration to say I was literally "dying" to tell someone my secret, but all these symptoms were nevertheless disturbing; something had to give. The release I needed came for me while I was attending an out-of-town meeting with some non-NIH colleagues. A friend nonchalantly asked, "So how is it going at NIH?" Noticing a surge of adrenaline, I almost spilled the beans right then and there. I wanted to pour it all out and have it over with. I restrained myself, but since I knew I had only a few days to go before the announcement, I dropped a strong hint. I said, "Well, all I can say is that this is going to be a pretty big week," which I punctuated with a sly wink. Just this innocuous hint was sufficient to make me feel as if I'd turned a valve and let out some emotional pressure.

Later that week I made the formal announcement via a mass e-mail to my colleagues at NIH and around the country. Hitting the "send" button produced in me a feeling of great release and relief. And shortly after I'd made the announcement, my flulike symptoms disappeared. I was back to my old self.

My NIH story illustrates how relatively minor forms of concealment can have noticeable effects. But what happens to people who experience major life adversities and don't talk to others about them? Dr. James Pennebaker resolved in the early 1980s to answer this question. He was intrigued by studies that indicated the magnitude of undisclosed traumas. For example, in a study of childhood and adult stressors with 200 adults, he discovered that more than 30 of them had an undisclosed childhood trauma.[19] There had been few well-designed studies that tested whether concealing traumas was actually health damaging, so he set out to put this concept to the test.

Pennebaker believed that the nature of some extremely negative life events either facilitated or inhibited open discussion about them. He was particularly interested in how people whose spouses had been accidentally killed would compare to those whose spouses had committed suicide. The reason? Pennebaker hypothesized that, because suicidal death is less socially acceptable than death by accident, spouses of suicide victims would be less inclined to talk to others about it. He believed that the stigma of suicide would lead spouses to conceal their thoughts and feelings about the death, and that this concealment would eventually cause more illness in this group than among the spouses of accident victims.

To test this hypothesis, Pennebaker and his student Robin O'Heeron surveyed spouses of suicide and accidental-death victims years after the deaths, asking them about their own health.[20] As he expected, health problems in all of the spouses increased the year after the death (see Chapter 13 on emotions for more on bereavement and health). Contrary to expectations, there was no statistical difference in health problems between the two groups. Surprisingly, the spouses of suicide victims were, if anything, slightly healthier. About these counterintuitive findings Pennebaker writes, "As Aldous Huxley noted, this appeared to be a case of 'the slaying of a beautiful hypothesis by an ugly fact.'"[21]

But what Pennebaker and O'Heeron *did* find was even more fascinating. Regardless of why an individual died, the more that the surviving spouse discussed the death with friends, the fewer health problems he or she experienced. Ruminating about the death was also predictive of health outcomes. Rumination is a vicious cycle. In essence it is the repetitive dwelling on a problem, with a focus on how bad you feel and how awful things are, without taking corrective action to make things better. Researchers have discovered that rumination tends to extend the period of time people are sad or anxious.[22] Rumination, if it continues unabated, can make a bad problem even worse. Pennebaker and O'Heeron found that the less the surviving spouses ruminated about the death of the loved ones, the healthier they were. And there was also a correlation between ruminating about the death and talking about it to others. Those spouses who talked about the death more ruminated less.

Pennebaker told me that the findings both surprised and relieved him. "We had predicted that spouses of individuals who had committed suicide would be less likely to talk about it and that they would have more health problems than [would] spouses of people who had died in car accidents. On one level we were wrong: Spouses of suicides talked much more about the deaths than [did] spouses of motor vehicle accidents. Nevertheless, the talking-health link proved to be correct."

Talking about it seemed to help the study participants move beyond their bereavement. Those who did not talk to others about the death might have been trying to get beyond it by avoiding it. But such avoidance, as you read earlier, can backfire, setting off a tug-of-war between the operator and monitor that increases access to the very thought one is trying to avoid. The spouses may have activated their operators to help them avoid thoughts of their deceased spouse, only to have their monitors bring such thoughts back to their awareness. Under distress of this type, attempting to completely avoid troubling thoughts only accentuates distress through the natural, and most often helpful, give-and-take of the operator and monitor.

For psychologist Pennebaker, the study of the bereaved spouses was revelatory, and it led him to begin a program of research that has produced some of the most intriguing, stimulating, useful, and often surprising findings in our field.

Chapter 5

Emotional Disclosure: The Remarkable Benefits of Opening Up

Matt Varney did not think he was having a problem opening up after the shooting at his high school, Columbine High, in Littleton, Colorado. This young athlete's life might have been spared that fateful day because of a timely invitation from a teammate to have lunch off campus, and during the aftermath he was talking to people all the time. He explained, "I became a spokesperson for Columbine, doing as many things as I could get my hands on." Varney was a perfect candidate for media interviews because he is bright, articulate, mature beyond his years, and not nervous talking in front of groups.

He says that he grew up quickly and became an outspoken example of healing and recovery. He gave heartrending speeches and wrote emotion-filled essays. At the one-year anniversary of the shooting, he gave a speech that was broadcast across the globe.

He thought he was doing the right things, but when his friend Greg Barnes hanged himself just two weeks before graduation, Varney realized he'd been glossing over his emotions. He says he had engaged in "healing backwards"—presenting an image of composure, of someone who had dealt with the horror of the tragedy and found some level of acceptance. But Varney said the truth was that he had "run from the emotion for an entire year."

After Greg's death Varney's emotions "started to pour out," and he had to face them. He explained, "I had given speeches about the importance of kindness, communication, friendship, and love. I put a lot of pride into how I had accomplished those things in my life. But I failed to deal with the pain in my life."

When Matt retraced his steps to healing, he included disclosing his feelings to friends and family. "When I finally broke down and told my own stories to others, it relieved me from having to deal with it all by myself. The loving support of [others] could not erase the pain of my loss, but they encouraged me to persevere. I believe that when we tell our stories, it helps articulate our new understandings."[1]

Is there something inherent in the expression of emotion, in confession, or in talking that is health enhancing? After his initial study, Pennebaker wondered. To find out, he and his colleagues launched a series of studies that are now scientific classics. Their studies, along with those of other scientists, have irrevocably altered and expanded our understanding of the importance of disclosure.

In these groundbreaking studies, Pennebaker developed an approach now widely used for eliciting disclosure in a confidential way from participants about traumatic events or stress. It involves having the participant visit the laboratory to write about the trauma (or an assigned trivial topic, for comparison) for a fixed amount of time, usually fifteen to thirty minutes, over the course of several days, ranging from one to four. Psychological, biological, and health assessments are made before and after the writing sessions, and again sometimes up to six months later. Participants assigned to write about trauma or stressful experiences are given instructions similar to the following:

Once you are escorted into the writing cubicle and the door is closed, I want you to write continuously about the most upsetting or traumatic experience of your entire life. Don't worry about grammar, spelling, or sentence structure. In your writing, I want you to discuss your deepest thoughts and feelings about the experience. You can write about anything you want. But whatever you choose, it should be something that has affected you very deeply. Ideally, it should be about something you have not talked about with others in detail. It is critical, however, that you let yourself go and touch those deepest emotions and thoughts that you have. In other words, write about what happened and how you felt about it, and how you feel about it now. Finally, you can write on different traumas during each session or the same one over the entire study. Your choice of trauma for each session is entirely up to you.

In the first study of the effects of disclosure, Pennebaker and his student Sandra Beall randomly assigned college students to write about a superficial topic or a traumatic experience for fifteen minutes on each of four consecutive days. Participants writing about traumas were further divided into three groups: (1) those asked just to vent their emotions about the trauma but not to mention the trauma itself (trauma-emotion group), (2) those asked just to write about the facts surrounding the trauma (trauma-fact group), and (3) those asked to write about their deepest thoughts and feelings and about the facts surrounding the trauma (trauma-combination group).

The college students wrote about gripping traumas. Pennebaker describes one student who, at age ten, did not follow her mother's instructions to pick up her toys before her grandmother's visit. The grandmother slipped on one of her toys, broke her hip, and died during hip surgery. The student blamed herself for her grandmother's death. Other essays were filled with stories of sexual abuse, alcoholism, concealed homosexuality, bereavement, public humiliation, and suicide attempts.[2] Pennebaker was taken with the gut-wrenching nature of their experiences and the candor with which these experiences were shared. Pennebaker writes, "The grim irony is that, by and large, these were eighteen-year-old kids attending an upper-middle-class college with above-average high school grades and college board scores. These are the people who are portrayed as growing up in the bubble of financial security and suburban tranquillity. What must it portend for those brought up in more hostile environments?"[3]

Not surprisingly, writing about traumas was emotionally painful for the participants. On average the participants in the trauma groups reported increases in negative moods and had higher blood-pressure levels during the writing sessions compared to those who wrote about superficial matters. People typically feel worse immediately after writing about hurtful events in their lives. Over time, however, the feelings improve dramatically. Having an undisclosed trauma is in some respects like having a painful, infected tooth that you just try to ignore. Much of the time you can distract yourself from it, but it is always there in the background, tainting life's pleasures. With the tooth, an extraction or a root canal will produce more pain in the short run, but will be the first step toward healing. Pain—psychic and emotional—is also an initial accompaniment to writing about traumatic experiences. As with an infected tooth, you are opening wounds and raising issues that have long been buried and which you may not wish to disturb. But, fortunately, things do get better.

The findings from the study were amazing and showed a pattern that would be consistent across most future studies. Pennebaker and Beall discovered that six months after the writing sessions were over, the trauma groups reported more positive moods and fewer illnesses than did the comparison group. These effects were most pronounced in the trauma-emotion and trauma-combination groups (i.e., those who vented emotions without mentioning the trauma and those who wrote about both the emotions and the trauma). The researchers discovered something truly interesting when they reviewed medical records from the student health center. Before the essay writing began, the groups did not differ in their use of campus health services. But six months later the trauma-combination group had a significant drop in visits to the health center.

In his wonderful book, *Opening Up: The Healing Power of Expressing Emotions,* Pennebaker recounts his feeling upon seeing these results:

> I'll never forget the initial thrill of finding that writing about traumas affected physical health. But the thrill was tempered with a little anxiety. For every question that the experiment had answered, a dozen more questions appeared. Perhaps the most basic issue that haunted me concerned the trustworthiness of these findings. Were the effects real? Does writing about traumas really affect physical health? Perhaps we had just affected people's decisions to visit the student health center. Or even worse, maybe the findings were simply due to chance. . . . Being impatient, I had to know if we were dealing with something real.

Pennebaker's concerns were unfounded. His findings were very real and, as you will see, only a beginning.

GIVING VOICE TO TRAUMA, OPENING UP TO HEALING

Jonetta Rose Barras is an accomplished author, journalist, and poet who has experienced more than her share of trauma and stressful life events. She writes a convincing, poignant, and sometimes shocking account in *Whatever Happened to Daddy's Little Girl? The Impact of Fatherlessness on Black Women* of the emotional wound she sustained because of what she calls "father deprivation." She says in her book that she had suffered the anxiety and pain of losing a father three times before she was eight years old. Barras, who wrote a column for the *Washington Times* and whose writing has appeared in publications such as the *Washington Post, USA Today,* and the *New Republic,* availed herself of many resources to overcome obstacles, achieve success, and create a life for herself and her daughter. However, in her search for the love that she said she never received as a child, she stumbled in various relationships, becoming pregnant early, coping with the death of her first child, losing custody of her second child, getting divorced, and being battered by the men in her life. She described herself to me as the black sheep of the family, who spent an inordinate amount of time alone because she had very low self-esteem and felt completely unworthy.

Barras told me that when she was a child, writing helped her cope by allowing her to create characters imbued with the emotional characteristics that she yearned for in her life: "Very early, what helped me get through some of that was that I did like to write and read, and I wrote stories that allowed me to express emotions where my characters were more loving than I thought the people around me were." Through her writing she nurtured herself and filled emotional voids. She said, "I created within my little stories and my little poems a world where I felt appreciated and loved. So many of my earlier writings dealt with love, with being part of a group, a family, a relationship." Maturing as a person and a writer, she grew to appreciate the value of exploring her deep feelings in her writing. She explained: "The ability to express emotions in my writing has always been sort of my saving grace, because I've been able to express . . . not only my own personal pain but also my anger, disgust, and dissatisfaction with things that I see. It plays a big part in my life. It keeps me from getting depressed. It keeps me hopeful. So much of my personal writing is filled with emotion, and I put a lot of emotion in my journalism. After what I have been through, if I didn't have the opportunity to write, I would not be as hopeful about life as I am."

In contrast to Barras's need to express her feelings, Henri Landwirth is equally adamant about keeping his hidden. Landwirth is an award-winning philanthropist who is the founder of the internationally acclaimed Give Kids the World Village in Florida, a fifty-one-acre, nonprofit resort for children with life-threatening illnesses whose wish is to visit Central Florida's best-loved attractions, such as Disney World, Sea World, Universal Studios, and Wet 'n Wild. Give Kids the World Village provides tickets to the resorts and weeklong accommodations for the children, their siblings, and their parents at the fantasyland village. It partners with more than three hundred international wish-granting organizations who identify the children and provide transportation to Florida. By 1996, Give Kids the World had served more than twenty-seven thousand children and their families, and, thanks to its many corporate sponsors, the families pay nothing. Many things are special about Give Kids the World Village. One is that every day when the family returns from a day at the theme parks, each child and all siblings receive a toy. This comes to over sixty thousand toys a year!

You may wonder what motivated Landwirth to create this one-of-a-kind escape for sick children. He explains that he is giving kids something that was taken from his childhood—true happiness. At the age of thirteen, Landwirth was taken from his family, and he lived for five years in a series of Nazi camps. As might be expected, his experiences were unimaginably horrible. Every day he witnessed murder, torture, starvation, sickness, unrelenting mourning, and despair. Of his transfer from a labor camp to a death camp he writes, "We arrive at Mauthausen dressed only in rags. There, they took all our clothes from us. Everything. They stripped us naked and left us in the freezing weather for more than a day. This was a death camp. A terrible place. The Germans wanted as many to die by the weather as possible. And many did die. More than I can count or remember. There was so little regard for the life of human beings. A person mattered less than a piece of dust." But through a succession of what he describes as miracles, he survived the camps and ultimately made his way to the United States, where he achieved great financial success as a businessman.

What makes Landwirth's story even more fascinating and relevant for the topic of disclosure is that once he made it out of the camp, he never talked to anyone about his experiences. Not a word for fifty years. Not even to his twin sister, herself a concentration-camp survivor, whom he miraculously found after the war. These two extraordinarily close twins, both of whom experienced unthinkable horrors in the camps, never discussed their time there with each other. In his moving autobiography, he writes:

Remembering a painful past is one of the most difficult things a person can do, and I am no different. Most of the events detailed here, I have never before discussed with anyone, not even my sister, Margot, herself a survivor of the Holocaust. My sister and I survived the death camps where millions of our fellow human beings, including our parents, were murdered. It has taken me a lifetime, more than fifty years, to tell this story. I've searched deep within my own being to answer why I did not speak of these events earlier in my life. While there are many reasons, self-preservation and protection of my sanity are two of the primary ones. The odyssey of my life has now brought me to a place in time when the reasons for silence are no longer compelling enough to remain quiet.[4]

Apparently Landwirth is not alone in his relative silence about his Holocaust experience. Pennebaker, who has worked with many of these survivors, reports that only about 30 percent have discussed their experiences in the camps after coming to the United States. Their reasons were many, and included wanting to move on, not wanting to upset their children, or thinking no one would understand.[5] But their silence did not mean that they were without daily thoughts of the Holocaust. One woman recounted to Pennebaker during a study the following haunting and recurring image:

They were throwing babies from the second floor of the orphanage. I can still see the pools of blood, the screams, and the thuds of their bodies. I just stood there afraid to move. The Nazi soldiers faced us with their guns.[6]

Working with the Dallas Memorial Center for Holocaust Studies, Pennebaker and his colleagues videotaped interviews with more than sixty survivors while taking their physiological measurements. From the content of the videotapes and the physiological data, they classified each survivor as a "high discloser," "midlevel discloser," or a "low discloser." A year after the interview, they discovered that the high and midlevel disclosers were significantly healthier than they had been the year before the interviews. This better health in the year following disclosure occurred despite the fact that all survivors were beyond sixty-five years of age. Conversely, the low disclosers were much more likely than the high disclosers to have visited a physician in the year after the interview than in the year before the interview.[7]

Pennebaker's work was able to scientifically tap in to the experiences of people like Barras, Landwirth, and many others. It also opened the floodgates of research on the effects of disclosing one's deepest thoughts and feelings about past traumas. Not all the studies used the essay-writing approach; in some, participants talked into a tape recorder. The results are similar with either method, although writing has the advantage of convenient tools (paper and pencil) and no need for quiet or privacy as required while speaking into a recorder. In some studies, participants wrote or talked about a variety of topics, including the process of disclosing itself. In other studies, all participants were assigned to make disclosures about one life experience such as a job loss, the stress of attending college, or even surviving the Holocaust.

Here are some of the results that have been discovered:

Disclosure improves mood. When people disclose their deepest thoughts and feelings about stressful or traumatic experiences, negative emotions such as depression and anxiety eventually lessen.[8]

Disclosure improves perceived health. After participating in the essay-writing procedure, people reported fewer physical symptoms such as upset stomach, headache, or racing heart. They also reported fewer visits to physicians and better coping with existing illnesses.[9]

Disclosure decreases doctor visits. A more objective measure of improved health is the number of visits participants make to physicians. Following trauma-related essay writing or tape recording, physician visits have decreased by as much as 50 percent and lasted for as long as one year.[10]

Disclosure improves objective measures of health in medical patients. Participants in most studies of disclosure are relatively healthy. The Smyth study described earlier showed that emotional disclosure can improve the health of people with asthma and rheumatoid arthritis. When the results from all patients were combined, it was discovered that 47 percent of the patients who disclosed stressful events showed clinically relevant improvement, whereas only 24 percent of the control group exhibited such improvement.[11]

Disclosure boosts the immune system. Several studies have found that shortly after episodes of emotional disclosure, participants show a variety of improvements in their immune systems, including higher levels of natural killer cells and CD4 cells and an enhanced immune response to the Epstein-Barr virus and to the hepatitis B vaccination (i.e., disclosure improved the effectiveness of the vaccine).[12] Suppressing thoughts about traumas, on the other hand, can cause a decrease in immune functioning.[13]

Disclosure affects reemployment following job loss. Another striking example of the effects of disclosure was discovered in a study of men who were laid off from their jobs and who for four months were unable to find new employment. In the study a portion of the men wrote for thirty minutes a day for five consecutive days about their deepest thoughts and feelings about losing their jobs. Another group wrote for the same amount of time, but about how they'd been spending their time since the layoff. A third group of men did not write at all. All men went on the same number of interviews. Amazingly, 27 percent of the men who wrote about being laid off found jobs within the next three months, whereas the reemployment rate for the other groups was only 5 percent. Several months later the differences were even more striking: By then 53 percent of the emotional disclosure group had jobs, versus only 18 percent of the other groups.[14]

Pennebaker believes that the differences had to do with how the men dealt with their considerable anger for their former employers. He writes, "These men felt betrayed by their previous employer. Even during our initial interviews, we often found it difficult to stop the men from venting their anger. We suspect that when most of them went on interviews for new jobs, many would let down their guard and talk about how they were treated unfairly. Those who had written about their thoughts and feelings, on the other hand, were more likely to have come to terms with getting laid off and, in the interview, come across as less hostile, more promising candidates."[15]

Disclosure improves grade-point average. In addition to improving health and emotional well-being, the act of disclosing your deepest feelings and thoughts about stressful life events can be beneficial in other ways. Research demonstrates that students who write in depth about traumas show greater improvements in grade-point average compared to students who do not disclose traumas.[16]

WHY DOES DISCLOSURE WORK?

Margie Levine said in an interview that she comes from a long line of women survivors: "My mom was feisty and dealt with a lot of adversity." This turned out to be a priceless legacy. Levine found herself struggling to stay alive, in the same hospital and at the same time that her mother was there recovering from kidney surgery. Determined to keep this cataclysmic change in her own health status from her mother, Levine covered her hospital gown with her trench coat when she visited her mother's hospital room. When Levine heard that her own condition was terminal, she prayed only to live long enough to prevent her mother from suffering the loss of her child.

For weeks before their eventual simultaneous hospitalizations, Levine had not been feeling well, but she ignored warning signs to plan her mother's care. However, one day, en route to visit her mother, she experienced a stabbing pain in her chest that would not be denied. She asked her mother's doctor to arrange an X ray, which revealed a large mass near Levine's heart. She was to learn that she had an asbestos-related form of lung cancer—malignant pleural mesothe-lioma, which kills in months.

Levine was a health-education teacher and a social worker before her illness, and she used all her training to marshal the forces that contributed to what can only be called a miraculous recovery. She convinced her doctors to try a radical treatment plan and incorporated nontraditional therapies as well. Levine says of her treatment, "The treatment was so horrifically aggressive. I was so sick. Radiation, chemotherapy, and three surgeries within six weeks." It has been more than thirteen years since she waged war against her disease, and she may be the world's longest-living survivor of this deadly cancer. Levine was so traumatized by her ordeal that she didn't talk to anyone but close friends about it for nine years. She understands that concealing trauma is not good. "There is a piece that's toxic," she says. "It's harmful to keep a secret, if you totally keep a secret, but if you share it with loved ones, it's not."[17] Before Levine could breathe a word to anyone, she started to write about what was happening to her. In her book *Surviving Cancer,* which has become a kind of guiding light and owner's manual for people suffering from terminal illness, Levine reveals, "I purchased a purple-lined notebook and pen [on the way home from Sloan-Kettering]. I knew then that I needed to tell my story, even before it had unfolded. I wrote and cried all the way home."

"I wrote ferociously, describing my fear, frustration, and struggles. . . . When the weight of illness seemed almost too great to bear, my journal was like a dear friend, available anytime day or night, a friend who would not turn away no matter how blue or downhearted I became. Here in the privacy of a leather-bound book, I could be raw and candid. There was no holding back. I could examine my complex emotions; I could grieve for the old life I wanted to be living; I could find refuge from the consuming pain, nausea, and fear, and it was here where I found courage to raise my sails again. Journaling not only gave me an important outlet for venting feelings safely, it also generated important insights that enhanced my healing."[18]

Dr. Louise DeSalvo, a professor in the Department of English at Hunter College in New York, is a successful author and a celebrated memoirist. She turned her pain into powerful prose in *Breathless,* which relates her experiences with adult-onset asthma; in *Vertigo*, which recounts her girlhood experiences with depression; and in *Adultery*, which describes her husband's affair when he was a young medical resident.

In 1999 DeSalvo incorporated research with her firsthand experience to create the text *Writing as a Way of Healing*, having written her own journals since 1981. During an interview she explained that she read Pennebaker's *Opening Up* as soon as it was published in 1990 but tucked the information away, not using it again until she developed asthma. She explained, "I had asthma. I wrote *Vertigo*, and the asthma got better simply by [my] writing *Vertigo*. That's when I started paying attention." She combined writing and psychotherapy to access healing. "I was really very desperate, and I figured I would use everything I could get my hands on to help, because I was so disabled," she says. "The therapeutic alliance certainly forces you to look at things in the same way that Pennebaker does. I have had insights in my writing that I have never had psychotherapeutically. There are things, of course, that psychotherapy can do that writing cannot. But I have been, on my own, able to get to things and then bring that into the psychotherapeutic alliance."[19]

How is it that the act of disclosing our deepest thoughts and feelings about stressful or traumatic life experiences produces these dramatic results? You would probably guess that it has something to do with allowing people to express or vent emotions—maybe the cathartic effect of releasing pent-up feelings? But studies suggest something different. When people write superficially about their emotions, without deeply exploring their thoughts and feelings, they do not show benefits. And the experience of catharsis is neither necessary nor sufficient to produce benefits of writing deeply; the benefits can occur with or without catharsis, but never without writing deeply.

Maybe you think it has to do with the possibility that disclosure somehow facilitates healthier behaviors? Well, it doesn't. Maybe it has to do with improving social relationships, which, you will read in the next chapter, are themselves health enhancing? No on that one, too.[20]

So why does it work? No one knows for sure, even the patriarch of this field, Pennebaker, who says, "Despite the dozens of studies, we still aren't certain why this writing technique is so helpful. My intuition—backed up by some research—suggests that a number of factors are at play." Here are a few of the reasons disclosure might work:

Disclosure disinhibits. One way emotional disclosure might help is that it counteracts the negative effects of thought suppression, emotional inhibition, and concealment that I described earlier. But as is often the case in science, things are not always as straightforward as one might assume. In fact, some research suggests that the benefits of disclosing trauma may not have as much to do with disinhibition as once thought. For example, Drs. Melanie Greenberg, Camille Wortman, and Arthur Stone of SUNY at Stony Brook conducted a study in which college students were asked to write about traumas that had never happened to them. The students writing about the imaginary traumas still showed significant improvement in health, as measured by visits to their private physicians and the student health center.[21]

Disclosure changes thinking and language. Research indicates that when people write in-depth about traumatic experiences, they begin to think differently about the event and about themselves. The greatest benefits from writing are seen in individuals who show an increase in the use of "causal" and "insight" words. Causal words include ones such as "because," "cause," and "reason." Insight words include "know," "realize," and "understand."[22] Why would an increase in the use of such words be helpful? According to Pennebaker, "Causal words are a reflection of the way a person is thinking about the event, indicating perhaps a greater sense of understanding or coherence. If I can tell you about a trauma and also give you a sense of why it happened and how it affected me in the long run, then I probably have a better sense of it than if I can only describe what happened, without giving a *reason* for why it happened. Causal words aren't important in and of themselves. Rather, they are useful as signs that a person is thinking more deeply about a trauma."

Disclosure helps develop meaningful stories. Pennebaker also believes that disclosure, especially through writing, helps you construct meaningful stories about your experiences, deepening your understanding of what you've gone through. It's as if causal words help a person construct a meaningful story. "I think when we can make a coherent story about an upsetting experience, we are able to move on," notes Pennebaker. "The mind naturally tries to understand complicated events, and, if we don't understand them, we keep playing them over and over in our heads. A coherent story breaks this cycle. But the healing effects don't stop there. Once we can move past the events, our social worlds will often change. We can talk and listen to our friends better. We might see that some of our old friends weren't helping us in the ways we needed help. And so telling our stories may be affecting how we think, how we feel, and how we interact with others." In an interesting test of the meaningful-story hypothesis, Joshua Smyth had study participants write about traumas in either a narrative expressive style (story form) or a fragmented style, where they simply listed their thoughts and feelings in bullet form.[23] According to Smyth, "Both groups expressed more thought and emotions about trauma than did a control writing group, but only the narrative group showed any evidence of improvement. This suggests that the organizational aspects of narrative storytelling appeared to be critical to improvement."

Disclosure in Action

Although Linda Ellerbee, the famed journalist mentioned in chapter 4, had the unconditional support of friends and family when she was diagnosed with breast cancer, she told me that she turned also to her writing: "I have found the process of writing about my experiences helpful all my life, and I have kept journals all my life. Sometimes I don't know what I think until I see what I write. I would write my emotions in my journal, sometimes emotions I could tell no one else. You don't want to say to your loved ones over and over, 'I'm afraid of dying' or 'Last night I dreamed I was going to die.' They don't need that burden, but I needed to let it out of me. I needed to get it out of me."

Walter Anderson reaped the benefits of disclosure in an astonishing way. His is a name not as familiar as is *Parade*, the newspaper magazine that is synonymous with Sunday papers across the country. Yet Anderson is the award-winning journalist who edited the magazine for two decades, and he is greatly responsible for its popularity and far-reaching presence. Even those familiar with his success are stunned to learn that he was raised in extreme poverty in a household fathered by an illiterate, violent, abusive alcoholic who beat Walter whenever he caught him reading. Anderson recounted for me an experience he had after writing about his childhood experiences in his first published book, *Courage Is a Three-Letter Word*. He said, "I thought I had overcome my anger, but I hadn't. I have a lot of reason to be angry. I was forty years old when I wrote that, and I tell the story of my father and how I grew up. I did not realize what I had done until I was speaking before a group about a year after the book was published, and I was introduced by a professor who said the book should have been titled *I Forgive My Father*. Then he introduced me, and I could barely speak, because I realized he was right. Finally I had forgiven my father, and I was free because I had exorcised the anger by being able to discuss it, by being able to deal with it on an intellectual plane—not intellectualize it but bring it to the surface and talk about it and share it. It actually got rid of my anger."[24]

Some writers intentionally use their writing gifts to explore and resolve personal problems and issues. Take the poet Maxine Kumin's *Inside the Halo and Beyond*, the journal of her recovery from a nearly fatal horseback-riding accident. The "halo" refers to the head restraint she wore while the broken vertebrae in her neck healed. She begins with an account of the day of the accident and proceeds through her recovery, which includes her time in intensive care, her difficult rehabilitation, and dealing with depressive thoughts. She ends with what has been called a miraculous healing: she literally got back on the horse. Her daughter encouraged her to write the journal, which Kumin dictated to her. The journal obviously became a part of her recovery—as she says in the text, "everything comes pouring out."[25]

Terry McMillan is the wildly popular, critically acclaimed author of five novels: *Mama, Disappearing Acts, Waiting to Exhale, How Stella Got Her Groove Back*, and *A Day Late and a Dollar Short*. So far three of her novels have been made into successful feature films. To write novels about the lives of black women, McMillan culls from her own life experiences, which include coping with the death of an alcoholic father when she was a teenager, conquering substance abuse, and losing her best friend to cancer the year after her beloved mother died from an asthma attack in 1993. McMillan was devastated by grief, and her mother's death stalled the completion of her last novel, which features a character with asthma, very similar to her mother. McMillan had learned strength and resilience from her mother, and the healing she experienced during an extended trip to Jamaica became legendary in the novel and movie *How Stella Got Her Groove Back*. Celebrity provides no immunity from the pain of life, but McMillan's craft has provided her with a resource for good health and, perhaps, longevity. She acknowledges, "This writing stuff has saved me. It has become my way of responding to and dealing with things I find too disturbing or distressing or painful to handle in any other way. It's safe."

"Writing is my shelter. I don't hide behind the words; I use them to dig inside my heart to find the truth. Plus, writing seems to be the only way I've been able to garner any real control over a situation or at least try to understand it. I guess I can say, honestly, that writing also offers me a kind of patience I don't have in my ordinary day-to-day life. It makes me stop. It makes me take note. It affords me a kind of sanctuary that I can't get in my hurried and full-to-the-brim-with-activity life."[26]

WILL DISCLOSURE WORK FOR YOU?

It is natural that gifted professional writers would explore serious issues using their well-honed skills, but you don't have to be a Pulitzer Prize–winning writer to take advantage of the benefits of disclosure. Emotional disclosure is a powerful technique that seems to show no demographic preferences. It works across various age groups, from young adults to senior citizens; it works for both men and women; it works for college students and for prisoners with a sixth-grade education; and it works for people from different ethnic origins. It has produced comparable results in Belgium, Mexico City, the Netherlands, and New Zealand.[27]

Louise DeSalvo said that she knew intuitively that writing could be useful in healing and could have broad applications. "The reason I was so interested in *Writing as a Way of Healing* is in one sense because I teach, but also because of my class of origin. My parents were working class, and psychotherapy was a well-to-do person's luxury. What do you do when you can't do that? What do you do for yourself?" Completing her text, she says, is her way of "putting something out there that might be useful for people who need to unravel things but do not have the privilege of entering therapy."[28]

A key determinant of whether or not the technique will work for you is how you approach it. Disclosure is not just writing about emotions or the facts of a trauma. You have to go deep into your thoughts and feelings to reap the benefits. Pennebaker's classification of individuals as either high or low disclosers is based not on whether they experienced trauma or stress but on how they talked about it. When discussing their stressful experiences, high disclosers, compared to low disclosures, go into more depth and show more insight. They exhibit more emotion, even to the point of crying. They reveal more personal information, and they show more behavioral signs of stress, such as a wavering voice. Low disclosers simply look as if they're holding back their feelings, with a kind of matter-of-fact air, even when discussing traumas.[29] Low disclosers also select experiences to discuss that researchers consider less stressful and less personal than do high disclosers.[30] Perhaps as a consequence of this holding back, low disclosers do not obtain the health and biological benefits of disclosure.[31]

So what does this mean for you? Pennebaker recommends some basic guidelines you can follow:[32]

Pick a topic that you feel needs some resolution. It doesn't have to be the most stressful or traumatic event of your life; it just has to be important to you and something that perhaps would be difficult to express to others.

Set aside fifteen minutes to write continuously. When you write, do so without worry about grammar or spelling. And you don't have to write every day, just when you feel the need.

Try talking into a recorder. Research indicates that writing and talking into a tape recorder produce comparable results. But paper and pen are easier to access, and they can be used even in public places.

Be a high discloser. The true benefits of writing about stressful events come to those who explore their deepest thoughts and feelings. When you write, describe what happened, how you feel about it, and why you might feel that way.

Expect some negative feelings. Remember, when you descend into emotional depths, in the short run you may have more feelings that are negative. Such feelings may occur immediately after writing but do not last very long. Over time your emotional well-being should improve.

Science suggests these specific ways that writing can be used to address trauma, but that does not mean that writing may not be useful for other issues. In fact, in addition to self-help manuals such as DeSalvo's, there are programs across the country that deal with writing to address various issues. One of the most popular structured programs was developed by Ira Progoff, a devotee of Carl G. Jung, and is called the Intensive Journal Process. His son Jonathan Progoff, the director of Dialogue House in New York, the administrative headquarters for the program, explained in an interview, "This is a process that helps you deal with things so amorphous they are hard to grasp—your emotions, your experiences, your feelings and intuitions. It's a tool for self-development, and it uses writing to get out our thoughts so we can work with them, and the whole goal is to get in touch with your emotions and experiences."[33]

As DeSalvo suggests, writing is a healing tool with the potential to reach the masses. It has an appeal. She points out, "Writing is cheap." It's also easy. Maybe too easy. You can be fooled into thinking that all you need is paper and pen, but don't view writing as a panacea. Writing won't solve all your problems, especially those rooted in traumatic experiences. Don't use writing as a substitute for other actions that might be more appropriate for your circumstances. In particular, it isn't clear from the research if writing helps people who are suffering from serious emotional problems, such as depression or chronic anxiety. Explore all alternatives available. Writing should not be used as a substitute for seeking the assistance of a mental-health professional. You may find a therapist who will work with you as you explore particularly difficult topics through disclosure writing.

Conclusion to Part II

The fact that behavior, such as disclosive writing, is a critical element of the new definition of health should not be surprising. National surveys show that most of us recognize that behaviors such as eating a balanced diet, staying physically active, avoiding cigarettes, or wearing seat belts can increase our longevity. These well-known health-promoting behaviors are among the best demonstrations of the historical success of behavioral and social science. But the search has continued to expand our options for taking an active role in our health. The research on disclosure is a result of this continued search. We now know that a significant number of us experience trauma and extremely stressful circumstances that we never disclose to anyone and that we actively try to avoid even thinking about. We know that such inhibition is hard work, psychologically and physically. We know that over time inhibition leads to short-term biological changes and an increased likelihood of illness. But research also indicates that traumatic experiences need not be debilitating. We know that systematically confronting past undisclosed traumatic experiences can counter the effects of inhibition. When trauma occurs, giving voice to our thoughts and feelings may be the best medicine. But giving voice to trauma does not mean venting your feelings or otherwise getting something off your chest. To achieve the health benefits of emotional disclosure, you must go deeper. This is not to say, however, that less intensive forms of written and verbal expression are without benefit.

The research on emotional disclosure still has many unanswered questions. For instance: Does disclosure have positive long-term effects on chronic diseases such as diabetes and heart disease or mental illnesses such as depression or anxiety disorders? By what biological mechanisms does disclosure affect health? How do the effects of disclosure compare to those of the most effective psychotherapies for stress-related problems? Scientists on the forefront of the new health science are currently tackling these and other questions.

The next two parts of this book depart from the psychological and behavioral elements of health and move into the social realm. One of the most significant discoveries in the new health science is that our thoughts and actions do not occur in a vacuum and that the social context is a powerful determinant of how we think, feel, and act. The social context in which we live can exert a profound influence on how optimistic we are and the degree to which we are exposed to trauma. It can also be a significant determinant of our health, above and beyond its effects on thoughts and actions. The next two parts, respectively, describe two features of the social element of health—social relationships and socioeconomic position.

PART III

ENVIRONMENT AND RELATIONSHIPS: SOCIAL IMMUNITY

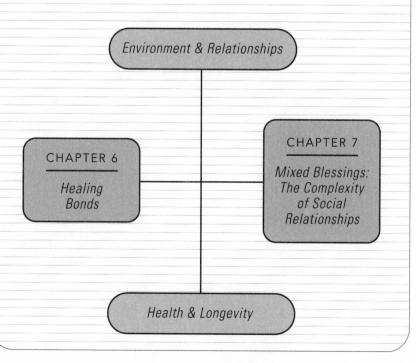

Environment & Relationships

CHAPTER 6

Healing Bonds

CHAPTER 7

Mixed Blessings: The Complexity of Social Relationships

Health & Longevity

A faithful friend is the medicine of life and immortality.
—Ecclesiasticus 6:16

Not all elements of the new and expanded definition of health are personal attributes. In fact, it has been recognized for some time that exposure to environmental factors such as lead paint, polluted air and water, pesticides, and hazardous wastes can have health-damaging effects. So much so that an entire institute of NIH, the National Institute of Environmental Health Sciences, is devoted to the study of these and other environmental risks. But there is another factor that resides outside the body that is also among the strongest predictors of illness and death. It is something that your body responds to in remarkable ways. It is your connection to others. When it comes to health and longevity, relationships matter.

Interest in the social origin of health and illness dates back to Hippocrates. However, until recently the medical establishment largely discounted the notion of a link between social relationships and health and illness. Now evidence from the social sciences is making this link hard to ignore. Through scientific findings and the stories of real people, the next two chapters show you that social bonds are more than emotionally fulfilling. Relationships can be life enhancing and lifesaving.

Think for a moment about how you would answer these questions:

- Do you have someone you can always count on to help you in an emergency?
- Is there someone you can confide in?
- Is your network of friends large or small?

- How often do you interact with family and friends?
- Are you lonely?
- Do you feel loved or appreciated?

In the chapters that follow, you will learn that how you answer questions such as these can affect your health and longevity as much as smoking and exercise. The power of relationships was illustrated most vividly for me during my mother's illness. When she was diagnosed with colon cancer, her physician estimated that she would live, at most, a few months. Yet nearly three years later she was still living and had been enjoying a good quality of life. Of course, predicting a person's life expectancy, even if that individual is very sick, is an inexact science, and physicians cannot be expected to be precise. But the fact that my mother lived *so much longer* than her prognosis is astonishing. I believe that one of the reasons is the fact that she was deeply embedded in a supportive social network.

As a minister in a large church, my mother always had many people who cared intensely about her well-being and who were ready to provide whatever assistance she needed. After her diagnosis, and especially when she became more fragile, the level of this caring increased exponentially. Cards, flowers, prayers offered in church, and phone calls all communicated one thing to her—you are loved and valued immensely. Beyond the overflowing well-wishes from the congregation, my mother benefited from the daily assistance of a smaller circle of friends, who formed a protective shell of tangible and nontangible support around her. These individuals furnished food, transportation, conversation, prayer, singing, quiet companionship, or whatever she needed that day.

Fortunately, I worked only fifty miles from her home, and, with the flexibility afforded me as a professor, I was able to reorganize my work life so I could spend as much time with her as possible. In fact, during her last three months I essentially moved back home, going to campus only to teach and attend meetings. One of my great joys was seeing my mother's face light up when, after she asked me, I told her that I would indeed spend another night with her.

Was my mother's unexpected longevity with cancer due to her being surrounded by a devoted and committed web of caring friends and family? Would others with a similar illness and with a similar support network also beat the odds? There is absolutely no way to answer these questions conclusively. A person's life span, whether that person is in perfect physical condition or battling a serious illness, is determined by many factors, some of which are biological, some behavioral, and some psychological. Rarely is just one thing responsible for illness and health. But social science has now demonstrated rather convincingly that *one of the things* that matters is our connection to other people.

Supportive social relationships operate in many respects like the immune system, which provides the body with continuous surveillance to prevent infections from taking hold and mounting an attack if invading substances gain a beachhead. The many dimensions of social relationships operate in a similar fashion, providing us with a level of ongoing protection against illness and often increasing their activity during personal emergencies. Supportive relationships provide us with a kind of "social immunity."

Chapter 6

Healing Bonds

A former colleague at Harvard, Dr. Lisa Berkman, is arguably the world's authority on the social aspects of health. She is chair of the Department of Health and Social Behavior at the Harvard School of Public Health, coeditor of the landmark text *Social Epidemiology*, and a member of the prestigious Institute of Medicine of the National Academy of Sciences. Interestingly, her distinguished career had its germination in the early 1970s, even before she entered graduate school.

Berkman was a young outreach worker with the San Francisco Department of Health, a job on the lowest rung of the public-health infrastructure. Pursuing a research career was the furthest thing from her mind, but by 1979 she had created a scientific revolution. Her job was to help community health centers respond to the health needs of their constituencies. To do this she traversed the working-class neighborhoods of San Francisco, talking to people on the streets and in their homes about their health concerns. Although she was successful in matching people with appropriate health services, the job was frustrating. "As an outreach worker you had very little opportunity to change any of the conditions in certain neighborhoods that seemed to be driving a lot of the health problems," she told me. "There was a mismatch between the services being offered and the larger needs of the community, in terms of things like jobs and housing."[1] Berkman believed even then that social conditions and health were somehow linked. Her desire to document that people's social environment might affect their health problems led her to enroll in a doctoral program in epidemiology at the prestigious School of Public Health at the

University of California at Berkeley. At the time, epidemiology and medicine were dominated by the *germ theory:* the idea that a specific disease has a specific cause—usually an agent such as a bacterium, a virus, or a microorganism. This theory presumes, for example, that you develop the flu from exposure to influenza viruses, tuberculosis from exposure to the tubercle bacillus, or lung cancer from toxins in cigarette smoke, and so on. According to germ theory,

A Specific Biological Agent ⟶ A Specific Biological Change ⟶ A Specific Illness

The germ theory presented a quandary for many scientists. They generally accepted the accuracy of the theory but felt there was more to the origin of illnesses than the one-agent-to-one-disease process. They recognized that everyone is constantly exposed to agents (viruses and bacteria), but many people never get infected, or they get infected but never get sick. Some scientists then hypothesized that whether or not an agent actually leads to an illness has something to do with a person's biological vulnerability, a concept called "host resistance." Essentially the theory says that certain agents can lower a person's overall resistance, opening the door for a variety of diseases and biological changes that can be pathological, not just one. The theory of host resistance was controversial in itself, but social epidemiologist Dr. John Cassel took it one step further. He proposed that one of the major determinants of whether individuals resist illness is their social environment.[2] According to Cassel,

The Social Environment ⟶ Alteration of Host Resistance (Vulnerability) ⟶ Illnesses or Health

Berkman was intrigued by the idea that social environments affect host resistance, because that was consistent with her anecdotal observations in the working-class or poorer San Francisco neighborhoods. For her doctoral dissertation she decided to explore whether people with larger social networks, which provide them with help and assistance, would have a lower-than-average likelihood of serious illness. To test this notion she selected the ultimate health concern: death. In an interview she told me, "In retrospect, I didn't recognize how radical it was to propose that something as general as social networks could be related to mortality." She found out just how radical it was during a meeting with her dissertation committee to defend her research proposal.

One of her committee members, a world-famous germ-theory scientist, voiced his strong opposition to the entire hypothesis that the social environment had anything to do with health. His words are forever etched in her memory, and Berkman recounted for me the story she has told many times before: "He said to me, 'Over the last hundred and fifty years of medical research, from Pasteur to Koch onward, research has proceeded successfully along the lines of identifying one cause of one disease, with the theory of disease specificity being one of the major advances in our thinking over the last century.' Then he asked me, 'Do you mean to tell me you believe that some vague concept like social forces could be a cause of disease?'" Whether it was confidence in her ideas or graduate-student naïveté, she answered then, as she would today, with an unequivocal yes.

Berkman's dissertation became landmark evidence on the relationship between social relationships and health, and it created a dramatic shift in the course of health-related social science and epidemiology. Although there was a good deal of research on environmental toxins, behavioral research on risks such as smoking and diet, that research still followed the germ-theory notion that specific biological changes, regardless of how produced, lead to specific illnesses. And the idea that the social environment could change host resistance and be health damaging was, at the time, quite revolutionary.

For her project she used data from the Alameda County Study, a nine-year study of more than 4,000 men and women in California, ages thirty to sixty-nine.[3] In 1965 all of the participants were asked about social ties in four areas: marital status, extent of contacts with friends and extended family, church membership, and involvement in other types of formal and informal groups. By combining these factors Berkman and her colleagues developed a measure of a person's "social network." As illustrated in Figure 3, they found that during the observation period of 1965–74, people in the study with a smaller social network *were twice as likely to die* than were people whose social network was larger. The effect of social network on survival persisted in a follow-up study of the surviving participants seventeen years after the initial assessment and was shown still to be as strong a predictor of death for those over and under the age of seventy.[4] Causes of death included heart disease, cancer, stroke, and gastrointestinal disorders.

It's important to note that the link between social network and survival was evident even after taking into account traditional risk factors such as age, gender, race, smoking status, health-care use, and physical activity. In fact, the size of the participants' social networks was a more powerful predictor of death than were more traditional risk factors.

Figure 3: Death Rates for Men and Women at Different Levels of Social Connections and at Different Ages (adapted from Berkman and Syme, 1979)

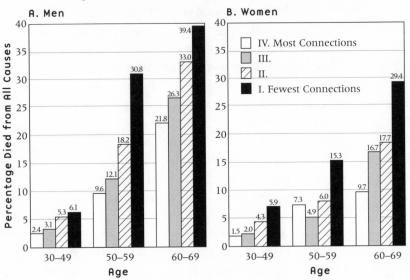

WHAT ARE HEALTH-ENHANCING SOCIAL RELATIONSHIPS?

Social relationships are as complex and varied as the human genome. They take many different forms and serve numerous functions. Some of us have a large collection of friends with whom we participate in a variety of activities. Some of us have only a few friends, but we enjoy very close and intense family ties. Others of us have a few people we can rely on in emergencies and a few confidants we turn to in times of extreme stress. Some of us feel loved and cared for by just one person in our lives. Unfortunately, some of us find that our social relationships are not supportive or fulfilling at all and are themselves a source of diminished well-being. Many of us are lonely, isolated, and cut off from meaningful interpersonal relationships.

When scientists want to measure social relationships, they devise methods for identifying the types and extent of each person's relationships. Two types of social relationships are *social networks* and *social support*. Social networks are defined in terms of the "structure" of our relationships, including the size, proximity, and amount of contact we have with friends and family. People who have an extremely small social network or fewer interactions with their network are said to be *socially isolated*. Those with larger networks or more contacts with their network are said to be *socially integrated*. Social support, on the other hand, is defined in terms of the "function" relationships serve for us, usually described as what one gives to or receives from another. Social support can come in several forms, including:

- **Emotional support:** help with emotional difficulties or upsetting situations or the communication of caring and concern
- **Instrumental support:** help with getting routine tasks accomplished, such as providing transportation, helping with chores, baby-sitting, or assistance in an emergency
- **Financial support:** help with economic needs
- **Appraisal support:** help with evaluating and interpreting situations or with problem solving

In addition to the above, researchers also examine relationships by asking whether a person:

- feels loved or cared for
- has a confidant
- is satisfied with his or her supportive social relationships
- is married or not
- feels lonely

Although among scientists there are subtle differences in the meaning of these terms and concepts, they all have one thing in common: Each has been linked, by well-controlled studies, to illness and death. Think about this: The effects of social relationships on cardiovascular disease are on *the same magnitude* as traditional risk factors such as cigarette smoking, high cholesterol, obesity, excessive alcohol consumption, and physical inactivity.

Those of us who rate our social connections as relatively high have, on average, more positive emotional lives, are more physically healthy, and live longer than those who rate them low.

Angie Hart's Network

Our friend Angie Hart and her family are at the high extreme of the social-network continuum. Angie has a large network of extended family that provides its members with continuous and emergency-based social immunity. Angie is in her late thirties, married, and the mother of four beautiful daughters, teens and preteens. She, her husband, Reggie, and the children live in a house on what my wife and I affectionately refer to as "The Compound." It's located in suburban Maryland, about thirty minutes outside of Washington, D.C.—close enough for convenience to the city but far enough for a pastoral environment. Four generations of Angie's immediate family live there on several acres of land in three houses that are in view of one another. Aunts, uncles, and cousins live nearby.

Angie lives next to her maternal grandfather, who shares a home with his wife and adult daughter. Behind those houses is Angie's mother's house, which we call "Command Central." This is the largest of the three houses and is the main gathering place for many extended-family social events. Angie's mother, June, built the house on land where she grew up, land that has been in her family for decades. It is a sprawling home with generous living quarters, spacious formal areas for entertaining, and smaller areas for intimate, informal gatherings. In addition to Angie's mother and stepfather, Angie's sister and brother and his wife and two children all have cohabited there comfortably.

Outside of The Compound, Angie and Reggie have an even larger extended family in the greater Washington area. Reggie grew up in Washington, D.C., where two of his three siblings still live, along with their spouses and children. With these combined families Angie has at least a hundred extended-family members living in and around the Washington area.

Relatives and friends gather regularly at June's house for Sunday dinners, birthdays, wedding showers, baby showers, receptions, anniversaries, parties, football games, or just to hang out. There's almost always something going on or being planned.

Obviously, Angie's social network is substantial, but she reports that it's not just the numbers. Angie says that this circle of family members provides nearly every kind of support imaginable, which I know comes mainly from first-degree family members with whom Angie has strong adult relationships. She feels emotionally close to these people, and there is a history of trust and reliability. The *informational* support, which Elizabeth and I turn to ourselves, is first rate. No matter what you need or need to know, someone in that family either knows or knows someone who does. They reduce the proverbial six degrees of separation down to maybe three. *Instrumental* support is also plentiful. For example, six children under the age of twelve have lived on The Compound at the same time. Angie's family really believes and lives the "it takes a village" idea, and everyone pitches in to provide child care. If Elizabeth and I have plans to go with Angie and Reggie to the movies, the theater, or dinner, there's always someone in the network who can watch over their children. Sometimes it's as easy as having the children walk down to "Gammie's," the children's name for June. It's also not unusual for Angie's children to be cared for by her brother, sister, or sister-in-law. And she often returns the favor.

The *emotional* support is unquestionable and unconditional. The concern for one another is always there, and everyone walks the talk. Although the family members easily could be completely enclosed and self-sufficient, they embrace the friendships of people outside the family, including Elizabeth and me. Each family member has a host of trustworthy, dependable friends. This enlarges the network, multiplying the potential sources of social support, further ensuring the social immunity of all those involved.

With all this social support available, you'd expect that longevity would also run in Angie's family, and it does. For example, Angie's grandfather is 86 years old, and his mother lived to be 104, with a good quality of life. Angie also has great-aunts who are in their eighties. None have debilitating chronic illnesses.

Of course, the health of Angie's family probably has as much to do with good genes as with good relationships. But there is no doubt that her family has *the added health benefit* of close social bonds, in addition to any inherited advantage. This is good news, since, while you cannot change your genes, you *can* improve the quality or quantity of your social ties. And, as I will discuss shortly, strong social relationships *can actually change biology* in ways that might increase longevity.

Healing Bonds: Discoveries from the New Health Science

In his autobiography, *The Measure of a Man,* Sidney Poitier talks about the closeness he felt within his family while growing up on Cat Island in the Bahamas:

> In the kind of place where I grew up, what's coming at you is the sound of the sea and the smell of the wind and your mama's voice and the voice of your dad and the craziness of your brothers and sister—and that's it. That's what you're dealing with when you're too young to really be counted into anything, when you're just listening, when you're watching the behavior of your siblings and of your

mom and dad, noting how they behave and how they attend to your feedings and how they care for you when you have a pain or when the wasp stings you around your eye. What occurs when something goes wrong is that someone reaches out, someone soothes, someone protects. And as the people around you talk, you begin to recognize things that are carried on the voice. Words and behavior begin to spell out something to you.

Poitier's reflections on his early life capture a feeling of caring and nurturance that has long been recognized as important to child development. Research is now establishing that a person's perception of parental warmth and closeness may have long-term health consequences.

A study of students at Harvard University in the 1950s confirmed that how we feel about our parents affects our health. Young men completed a series of physiological assessments and psychological tests, which included two multiple-choice questions about their feelings of warmth and closeness to their parents. One such question, as shown below, pertained to their mother, one to their father.

Would you describe your relationship to your mother [father] as (check one):

A. Very close
B. Warm and friendly
C. Tolerant
D. Strained and cold

A distinguishing feature of this study is that the researcher, Dr. Linda Russek, was able to locate nearly all the participants again in the 1980s, some thirty-five years after the original data collection. Again extensive medical and psychological assessments were administered, and participants' medical records reviewed. Dr. Russek and her colleague Dr. Gary Schwartz analyzed the data to determine whether the students' feelings of parental warmth and closeness in the 1950s had been predictive of their health as adults. Amazingly, *91 percent of the participants who had not earlier perceived themselves as having a warm relationship with their mother had been diagnosed with a medical illness in midlife, compared to only 45 percent of those who re-*

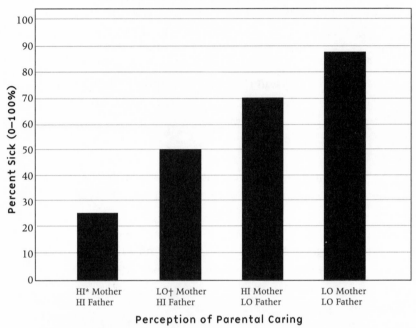

Figure 4: Percentage of Participants Sick at Midlife as a Function of Their Perception of Parental Caring 35 Years Earlier (adapted from Russek and Schwartz, 1997b)

*HI = high parental caring
† LO = low parental caring

ported a warm relationship with their mother. The illnesses included heart disease, ulcers, alcoholism, hypertension, and asthma. Perceived warmth and closeness to the father was also predictive of participants' health thirty-five years later.[5]

As illustrated in Figure 4, there was an additive quality to the perceptions of closeness to the parents. Those participants who felt both parents were caring were healthier later in life than were those who felt only one was caring, who were themselves healthier than participants who had felt neither parent was caring. In fact, *87 percent of participants who had felt neither parent was caring were sick at midlife compared to only 25 percent of those who felt both parents were caring.*[6]

These fascinating studies by Russek and Schwartz and by Berkman are examples of approximately twenty studies conducted around the world, involving more than 55,000 participants. Each demonstrates convincingly that people with high levels of social ties have longer life spans and lower rates of serious illness than do people who report lower levels of support. The strongest evidence to date is on the influence of social relationships on mortality. Following are several examples of these studies.

Relationships predict longevity. University of Michigan scientists studied more than 2,000 healthy men and women, ages thirty to sixty-nine, for nine to twelve years. A number of social relationships were measured, including marital status, church attendance, and participation in volunteer activities. *Women with smaller social networks had a mortality rate nearly double that of those high in social ties.* For men it was even more striking—*men low in social ties died at two to three times the rate of men with strong social connections.*[7]

At Duke University more than 300 elderly men and women were studied for thirty months to examine the effects of social relationships on mortality. Measures of social ties included how frequently participants interacted with friends and relatives, their social roles (e.g., spouse, parent), and their perception of the support received from others. *Those who were more socially integrated were between two and almost four times more likely to survive during the thirty-month study than were those less socially integrated.*[8]

Relationships predict heart-disease deaths. In a six-year study of more than 17,000 Swedish men and women, ages twenty-nine to seventy-four, researchers discovered that *the risk of dying from heart disease alone, as well as from all other causes, was 1.3 times higher among those with fewer interactions with their social network, compared to those with more interactions.*[9] In a five-year study of over 13,000 people in North Karelia, Finland, *men with fewer social connections were approximately 1.5 times more likely to die from heart disease and all other causes* than were those with more social bonds.[10]

Relationships predict heart-attack recovery. Approximately 190 elderly men and women who were hospitalized for heart attacks were studied at Yale University for six months after their attacks to determine survival rates. The results are shown in Figure 5. *During that time, those who lacked emotional support had nearly three times the death rate of those with emotional support. The data analysis took into account severity of disease, smoking, and other factors.*[11]

Relationships predict survival from coronary artery disease. A team at Duke University examined predictors of survival among more than 1,300 patients, primarily men, previously diagnosed with coronary artery disease.[12] The patients were classified based on whether or not they were married and whether or not they had a confidant—that is, someone they could trust and confide in. After five years Dr. Redford Williams and colleagues determined who was still living. The patients who were married, those with a confidant (whether married or not), and those with both a spouse and a confidant had similar survival rates. *Those who were unmarried and had no confidant had three times the death rate of the other groups,* after researchers accounted statistically for other risk factors.

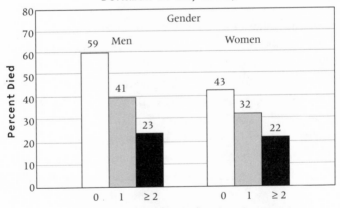

Figure 5: Relationship Between Emotional Support and Death (adapted from Berkman et al., 1992)

Relationships protect against the common cold. Dr. Sheldon Cohen of Carnegie Mellon University quarantined about 270 volunteers on a hospital ward, after assessing their levels of social involvement. During the quarantine period nasal drops containing a low dose of two infectious viruses known to cause colds and cold symptoms were administered. Before and after exposure to the viruses, participants were tested for any signs and symptoms of a cold, including respiratory symptoms such as congestion, runny nose, sneezing, cough, sore throat, chills, and headache, and any nasal congestion, discharge, and mucus production. In response to both viruses, the higher the density of social networks, the lower the susceptibility to colds, as shown in Figure 6. Compared to the group with a high level of social networks, *the group with the lowest level of social involvement was over four times more likely to develop colds and cold symptoms,* whether the symptoms were assessed by participant report or by objective biological criteria.[13]

Relationships predict progression of HIV to AIDS. Dr. Jane Leserman of the University of North Carolina at Chapel Hill studied 82 HIV-positive gay men for seven and a half years to examine social, psychological, and emotional predictors of AIDS. Assessments of AIDS or AIDS-related symptoms were conducted every six months. Men who reported less satisfaction with their level of social support at the beginning of the study were significantly more likely to develop AIDS than were men who reported higher satisfaction. In fact, *for every 1.5-point decrease in satisfaction with social support, the risk for AIDS approximately doubled.*[14]

Relationships predict birth outcomes. Social relationships not only affect our health and help us function in the world, they also help our mothers get us into the world. Researchers have discovered that pregnant women who are more socially integrated have fewer complications during childbirth and deliver babies who are generally healthier. Additionally, researchers in London and the United States assessed the social support of 247 pregnant women received from family, the baby's father, and other sources during pregnancy. They

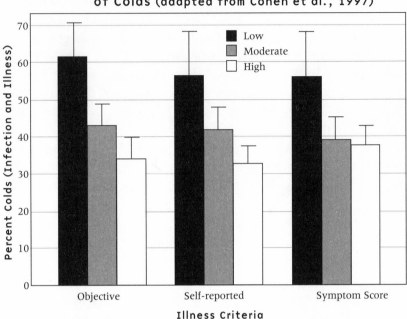

Figure 6: Social Network Diversity and Incidence of Colds (adapted from Cohen et al., 1997)

discovered that the growth of the fetus, as measured by birth weight, was significantly greater among women with higher levels of social support.[15]

The Love Connection: Spousal Support in Marriage

Before Diana Golden and Steve Brosnihan fell in love, she had lost her right leg, both breasts, her uterus, and most of her hope to cancer. But that had not stopped her from winning national and world championships in disabled ski competitions—including an Olympic gold medal in Calgary for the giant slalom—and becoming Female Skier of the Year in 1988.

Losing her leg never "handicapped" Golden—either on or off the snow. She triumphed over her disability with a courageous, optimistic zeal for life, retiring from competition in 1991 and becoming a much-sought-after motivational speaker. Then cancer delivered a one-two punch that sent her into despair. A biopsy revealed cancer first in her right breast, then her left, requiring a double mastectomy. Her resolve was weakening, and she had nothing left with which to face the removal of her uterus and the chemotherapy that followed.

Golden attempted suicide but reconsidered after taking an overdose of pills and called a friend, who rushed her to the hospital in time. Golden tried to fight back; she even bought a puppy, but it became sick and had to be euthanized, and Golden found herself in the Rocky Mountains planning a suicide jump into the Black Canyon. But the spirit that had propelled her as a twelve-year-old to ski six months after doctors had amputated her leg would not let her jump. Golden later decided to return to New England to be near her family for the remainder of her treatment. There, in Bristol, Rhode Island, at the Belcourt Castle, the snow princess grudgingly attended a Halloween costume ball, where she met her prince, who was dressed as a knight with a big fly's head—"a fly-by-knight." Brosnihan had designed the costume for himself, but nothing could have contradicted him more— a fly-by-night he was not.

He would not be dissuaded when Golden told him immediately about her cancer—and her terminal prognosis. Brosnihan had been admiring her vivacity from afar since their days at Dartmouth. He is quoted as saying, "Sometimes I'd see her crossing the college green on crutches, and I'd speed up just to get ahead of her, so I could see her smile." Now he was in love. A love that would change and maybe extend Golden's life. She admitted that, although she did not remember Brosnihan from college, she felt a special spark, too, but she was cautious about putting her emotions in harm's way. She is quoted as saying, "There was no way to expect such a thing could happen to me. I was worried. I wrote him a letter telling him to be careful. I said that if something else didn't work out in my life, it would kill me." Her heart was safe with Brosnihan. Their first date was at Dana-Farber Cancer Institute in Boston, where Golden was receiving chemotherapy. With Brosnihan there, she is also quoted as saying, "It was the beginning of a whole change in how I approached treatment." Brosnihan proposed the next Valentine's Day, while she was again receiving outpatient chemotherapy treatments. They married in August 1997.

Soul mates and life partners, Golden and Brosnihan turned her very frequent cancer treatments into events, arriving formally dressed in a limousine one New Year's Eve, staging practical jokes for the nurses, performing karaoke, or just playing Scrabble. Whether making chemotherapy "fun" or enjoying the quiet of their Rhode Island home, they made the most of every moment.

By all accounts, because of Brosnihan's love, Golden reclaimed her reason to live. With him by her side, she tolerated the grueling cancer treatments, and he helped her have a good quality of life. Golden said, "My life is very rich with people. Being in love and loving someone makes a big difference."[16]

The knight lost his snow princess when Golden died in August 2001. Yet without doubt Golden's relationship with Brosnihan helped her live better and perhaps longer than, at some points, even she thought she would.

One of the most intriguing contexts for getting and receiving social support is marriage. The health consequences of marriage have been studied for more than a hundred years, and the findings are consistent: Married people live longer and have fewer chronic diseases than do persons who are single, widowed, or divorced.[17] Even after developing a serious illness, married people fare better. A study of more than 27,000 cancer cases found that marital status was related to cancer progression.[18] Unmarried people were diagnosed when the disease was in a later stage; that is, the cancer had progressed further before it was detected. After accounting for stage and treatment of the cancer, unmarried people had poorer survival rates—they did not live as long as married people with similar cancer stages and treatment did.

The salutary effects of marriage are partly attributable to the nature of people who get married in the first place, what is known as *marriage selection*. People who marry are on average more mentally and physically healthy than those who do not marry, because single people who are very ill are less likely to marry.[19] Above and beyond the marriage-selection effect, there is also a *marriage protection* effect. Studies show that married people live longer even after a disease is present, suggesting that there's something more than selection at work. Although marriage is no panacea, being married somehow provides a level of protection against illness, disability, and death.

The protective effects of marriage are partly explained by economics. Married people have a higher socioeconomic position on average than do the unmarried, which is itself health protective, as you will read in a subsequent chapter. The economic benefits of marriage seem to be especially health protective for women,[20] although this may be changing given the recent economic strides achieved by women. Many experts believe, and the science confirms, that the health benefits of marriage are related more to the social support provided by the spouse.

University of Chicago sociologist Dr. Linda Waite, an expert in marriage and social support, told me, "Married people have a built-in source of social support. When the marriage is healthy, they have someone on their side, someone to talk to, and someone who cares. Having this kind of spousal presence in turn leads to other benefits, such as help in dealing with stress, help solving problems, and help in taking better care of oneself. This kind of spousal support just makes [people] feel better. And their health is better as a result. Social support is a key reason that married people live longer."

When it comes to the benefits of marriage, however, a big gender gap exists. Marriage has been found to benefit men's health more than women's. Specifically, although married people live longer than the unmarried, there is a larger gap in the life expectancy between married and unmarried men than between married and unmarried women. Why the difference? No one knows for sure, but it could have something to do with the fact that men report more overall satisfaction from their marriages than do women on average. Men place less priority on having wider interpersonal relationships than do women. Married men, it might be said, put all their social-support eggs into one basket—their wives.

The health-enhancing role that wives play for husbands was dramatically illustrated not long ago for Duke University basketball coach Michael Krzyzewski ("Coach K"). I followed the on-court fortunes of Coach K and his teams intensely during my fourteen years as a Duke professor. For some of my friends, the fact that I was a Duke basketball season ticket holder was more important than my being a Duke professor. Attending Duke games in the inimitable Cameron Indoor Stadium is as close to Nirvana as it gets for many college basketball fans. I was at Duke when Coach K built what is arguably the top college basketball program in the country. As of 2001 his teams have had seventeen National College Athletic Association (NCAA) tournament bids and have the highest winning percentage in NCAA tournament history. They've been to ten NCAA Final Four championships, of which they won three. They've won the only back-to-back NCAA championships in the last twenty-five years. Coach K is a six-time national Coach of the Year, and a five-time Atlantic Coast Conference Coach of the Year. Most of his players graduate on time. With all this success you might think that Coach K's career would be a complete joy for him. But in his book, *Leading with the Heart,* Coach K says that in January 1995, at the pinnacle of his career, he was ready to resign. He was emotionally spent.[21]

Coach K's crisis was the result of a pain problem that had progressed from bad to worse. It all began in the summer of 1994, with pain in his left leg from what he thought was something minor, perhaps a pulled hamstring. But the pain persisted for months, until he could barely walk. Eventually he was diagnosed with a ruptured disk and was scheduled for surgery in October, shortly after a new season of basketball practices began. Following a successful surgery, Coach K opted for a few days of recovery—instead of the ten weeks of limited activity the doctors recommended. Not a good decision. Over the next two months his condition deteriorated. He became weaker and began losing weight, and the back pain was worse than ever. Two days after enduring grueling plane trips to and from Hawaii for a tournament, he was barely able to get out of bed. That's when his wife, Mickie, stepped in and wanted to schedule a doctor's appointment for him immediately. Coach K protested that the appointment conflicted with basketball practice. Here is how Coach K describes what then transpired:

> "I'm setting up an appointment with the doctor and you need to be there," she said.

"I can't. I have to go to practice. I have appointments with the players. And then we're leaving for the Georgia Tech game. I don't have time."

"You don't have the strength," she said before leaving the room to call the doctor.

When she came back, Mickie informed me that the appointment was all set up for 2:30 that afternoon.

"But I told you that I have practice at 2:30!"

"Michael," she said, "I've never said this to you in our whole married life before, but it's me or basketball right now. If you don't show up at 2:30, I'll know what your choice was."

Not that it was easy for him, but for the first time in his entire coaching career, Coach K missed a practice.

And he admits that keeping that doctor's appointment was a turning point in his life. It not only led to a year away from basketball to heal, but also to a reorganization of his priorities and a more balanced life.

Coach K's back-pain episode demonstrates how spouses, especially wives, can step in with a kind of tough-love support at critical times. That does not mean that such support is always easy to accept. We expose ourselves to vulnerability when forced to rely on assistance from others—whether it's talking about a bothersome situation at work or help with a specific task. I have a friend and colleague whose wife told me that he'd been experiencing fairly severe chest pains but refused to see a doctor, even in the face of her repeated prompting. This despite having a family history of heart disease, a sedentary lifestyle, and a career at the Centers for Disease Control and Prevention! I sent him an e-mail berating him for not acting more wisely, to which he responded that his wife's entreaties had worked—he was finally setting a doctor's appointment.

These accounts typify what I refer to as the married man's pattern of social support. Married men report feeling more supported than single men. Why? Because they have spouses! Studies have shown that married men get nearly all their emotional support from their wives. Men feel that they receive more support from their wives than the wives feel they receive from their husbands. The loss of a spouse has more of a negative impact on the health of a surviving male partner than it does on a surviving female partner in a marriage. Men, with all their support coming from the spouse, tend not to have much to fall back on when that spouse is no longer available. Thus, the good news is that married men generally get health-enhancing social support from their spouses. The not-so-good news is married men don't do as good a job supplementing the support they get at home.

My own personal experience as a married man backs up the science perfectly. Before marrying I was very self-reliant, not seeking or receiving much substantial, instrumental, or emotional support at all. I had a few close friends with whom I communicated infrequently and numerous loose bonds with colleagues and pickup basketball partners. Generally I didn't realize that I was lacking anything. But after fifteen years of marriage, I see things much differently. There is a well-known saying: "I've been rich and I've been poor, and rich is better." Well, I've had low (or no) support and I've had high support, and high support is better. Elizabeth provides a good deal of instrumental support, showing me that I don't have to do everything for myself. She has also taught me that being more open to her about emotional topics not only makes me feel better but also makes our relationship stronger. Because of her encouragement my lifestyle is also healthier. I now get regular medical checkups, eat a healthier diet, take vitamins, do yoga, and take more time off work. Through her example I also see the value of friendships more clearly.

Elizabeth is always willing to provide assistance and knows instantly when something is bothering me (sometimes even before I know myself). I've learned that while I *can* handle the situation without my wife's help or support, it often works *better* when I access the support she offers. When we put our heads together to solve a problem, whether big or small, the outcome is usually one that I probably would not have arrived at on my own. If nothing else, I feel better for having aired the topic with someone who has my best interests at heart.

All this sounds great for men, but where does it leave women?

Fortunately, when it comes to the availability and use of social ties, women tend to have larger social networks than men do, and they receive support from a wider variety of sources. Elizabeth's network is a good example of this. She described it for me below:

Elizabeth's Social Network: Love on the Run

"My social network developed slowly, and the characteristics are constantly changing. My personality presents a kind of dichotomy, because I grew up painfully shy, but I love having friends. However, that early shyness gave me a deficit in the quantity category. I evolved into a caring and outgoing, yet private and cautious adult. Consequently, I have a small, carefully cultivated circle of friends. I love them intensely as they represent the small part of the world that knows me and still loves me. Since the numbers are small, a decrease in the ranks can have shattering effects, as Norman pointed out.

"My best friend, Brenda, started as a 'work friend.' We rarely socialized outside work, but saw each other every day, catching up on our family trials and tribulations over breaks and during luncheon outings. Neither of us came from close families of origin, and we began to fill essential dimensions of emotional support for one another. In fact, she was matron of honor when Norman and I married.

"When she died unexpectedly, I was devastated. I felt as if my *one* assigned friend was gone. I felt alone, like the Little Prince shown in the drawing on the cover of that children's tale. No one understood the depth and degree of my pain. Brenda had been my connection to the world. She validated me, cared for me, supported me, and helped so many things make sense. Her death left me feeling alienated and disconnected from the world, and I quite seriously thought I would live the rest of my life without another close girlfriend. Her death was like an emotional earthquake that had such an impact on my physical health that my menses stopped for six months. Because she and I shared a friendship that did not include our families, they did not appreciate what was happening to me. To make it worse, for several years after her death, I shielded myself from forming relationships with other women to avoid the pain of remembering Brenda or facing another loss.

"I believe that women place a high value on their female friends, especially those friends who have proven themselves trustworthy, loving, and reliable through bad times, bad boyfriends, and bad bosses. We provide lifelines to each other just by being together. We recognize in one another, and in each other's lives, similarities that enable us to endure (or enhance the enjoyment of) our own situations. I know older women who are shining a light on the path of womanhood for me and younger women who keep me connected to my youth while allowing me to develop my own powers of guidance. My opinion is that women do not fare well for very long without their women friends. Some of us realize this only after experiencing a loss as I did, but some of us know it from grade school. We make friends back then, and we hold on to them.

"I eventually developed close relationships with three other women who had also experienced the death of women friends. They are part of a small, supportive circle of about ten women friends whom I do not see regularly but each of whom is an essential component of my social network—and I hope I'm the same for them. We love each other from afar. We don't hang out much because of distance, children, work, or other commitments, but we know we can depend on one another, and we stay close and in touch regularly through e-mail and long-distance calls.

"I wish our interactions were more frequent, but it's the best some of us can hope for as we try to live in difficult times, follow dreams, and raise children. There is no substitute for an afternoon with a close friend, to laugh, shop, share secrets, give advice, gain perspective, and get a hug. However, we can ensure our emotional and physical health by finding creative, flexible ways to nurture our friends—to give and get what we need."

The Delany Sisters: Social Immunity and Longevity

The famous Delany sisters are a perfect illustration of ongoing social immunity that provided nearly every type of support imaginable. The relationship between Elizabeth (Bessie) and Sarah (Sadie) Delany may have helped keep them alive for more than a hundred years. Born into a family of ten close-knit siblings, these sisters formed an even stronger bond between the two of them that sustained them through the racism and sexism of the early 1900s. At the ages of 102 and 104 they chronicled their remarkable relationship in the bestseller *Having Our Say*. In the memoir Bessie wrote, "We were best friends from day one."[22] The companionship that began when Sadie, the elder, was just two years old provided the emotional support that likely contributed to their dazzling longevity. They lived together their entire lives, shunning marriage to pursue hard-won careers.

Not only did they live long, they lived well. Bessie was the second black woman licensed to practice dentistry in the state of New York. Sadie was the first black person to teach domestic science in New York City high schools.

The entire family was distinguished by its remarkable achievements. Each sibling had a professional career, spanning the fields of dentistry, science, music, and law. The bonds that developed as children surely contributed to their success in what can only be described as a hostile and repressive environment. In the book Sadie explained, "We all relied on each other. Throughout the years we lived in Harlem from World War I until Bessie and I moved to the Bronx after World War II, all of the brothers and sisters saw each other at least once a day."

The remarkable relationship between Bessie and Sadie gave them an edge that made them the only siblings to cross a century mark. Their coauthor, Amy Hill Hearth, explained, "They seemed to have conquered old age, or have come as close to it as anyone ever will. It was clear that they found the source of their vitality in each other's company. . . ."

Bessie was still climbing trees to prune branches when she was 98 years old, and Sadie authored *On My Own at 107* the year after Bessie died at the age of 104. Then Sadie joined Bessie again, dying at the age of 109 in 1999.

Safety in Numbers

Snow had fallen on greater Washington, D.C., helping set the mood for Christmas, which was just days away. Christmas at our friend Angie's is a wonderland even without the snow—decorations abound, presents are plentiful, and that elusive "joy of the season" really exists. The hum of Christmas was pierced by a call from Angie's grandfather to his daughter June, Angie's mother, to say that he was having chest pains. "Probably just indigestion," he was sure. June called Angie to come to her house right away to watch the grandchildren, who were visiting, so she could check on Granddad. Instead Angie went right over to Granddad's and found him still complaining of discomfort and still attributing it to indigestion. June, her sister Endia, and Grandad's wife joined Angie immediately. They asked him about foods that could have caused an upset stomach and assessed any symptoms that might suggest something more serious. Although he seemed to be doing better, the four relatives who had come to his aid put their heads together to decide what to do. Despite his claims of indigestion, they decided to call 911. This collaborative decision probably saved his life. Not five minutes after the emergency medical technicians arrived, while they were still taking his vital signs, Granddad's eyes rolled back and he lost consciousness. Angie's grandfather was in cardiac arrest. For the next few minutes, the emergency team worked to revive him, using a defibrillator and adrenaline shots to restart his heart. Their timely intervention was successful, as was his surgery later that day. On Christmas Day, five days after his heart attack, Angie's grandfather celebrated with thirty-five relatives and friends, who showed up at the hospital en masse for a visit. His life was likely saved by the well-developed family network of support and quick decision making. Her grandfather had people in his life who were available to him in a crisis and on whom he could rely, in this case for a lifesaving decision. Having a large social network can make a big difference for people suffering from chronic or terminal illnesses, especially those in need of substantial assistance. The demands of providing instrumental support in such cases can be physically and emotionally draining on even the most dedicated caregivers, who can burn out fast.

Lynn Mazur could see that the family of her friend Karen Hills, who was battling lymphoma, was in need of a respite. A cancer survivor herself, Mazur met Hills in a cancer-support group and was inspired by the book *Share the Care*[23] to organize a group of twenty to thirty family and friends to pitch in when and how they could to help Hills. This is a kind of business approach based on a project-management model to create what the authors call "funny families"—replacements for the extended biological families of another era. The secret is that no one has to do everything but everyone can help a little. Caregivers in the group benefit from the support and understanding of one another, feel less helpless because they have defined roles, and feel relieved because they have specific tasks to perform at specific times. In funny families people volunteer their expertise with whatever time available, knowing that their contribution, especially when added to that of the group, makes a significant difference. Mazur gathered people, based on a list from Hills, who could clean, shop, pay bills, run errands, help with Hills's physical therapy, and drive her child to school and after-school activities. Some people helped Hills with her journal writing, read to her, found devices that could make her more comfortable, or did whatever else she needed. Mazur made all the calls, set up a phone tree, and established "team captains" who checked with Hills's family each week to determine their needs and make assignments. The funny family was soon operating smoothly and became a "functioning extension of [Hills's] household." They infused Hills's family with energy and enabled them to interact more as a family during her final days. Mazur misses her friend deeply, of course, and regrets that Hills was never able to see the wheelchair ramp her funny family was building. Mazur's satisfaction is that her friend ". . . was able to go out on her own terms, knowing full well that she was loved." Hills's funny family is ready to go into action again to provide essential social support for someone else when needed. One of the team captains, Jackie St. Martin, said, "I would do it again without even thinking. I would do it for a friend of a friend. I would even do it for a stranger."[24]

PATHWAYS TO HEALTH: HOW DOES SOCIAL SUPPORT GET UNDER THE SKIN?

In 1994 Morrie Schwartz, made famous by the Mitch Albom book *Tuesdays with Morrie,* was diagnosed with Lou Gehrig's disease or, more technically, amyotrophic lateral sclerosis (ALS). The disease, which developed when he was in his seventies, began to slowly destroy his neurological system, ultimately rendering him incapable of using his body, while leaving his mind fully alert. In this excerpt from the book, Morrie explains the value of his social bonds.

> Say I was divorced, or living alone, or had no children. This disease—what I'm going through—would be so much harder. I'm not sure I could do it. Sure, people would come visit, friends, associates, but it's not the same as having someone who will not leave. It's not the same as having someone who you know has an eye on you, is watching you the whole time.
>
> This is part of what a family is about, not just love, but letting others know there's someone who is watching out for them. It's what I missed so much when my mother died [when he was a child]—what I call your "spiritual security"—knowing that your family will be there watching out for you. Nothing else will give you that. Not money. Not fame. . . . Not work.[25]

Now that we know that social support improves health, the question is, how? How is it that knowing someone is "watching out for you," as Morrie Schwartz says, can translate into a health benefit? One way is that social relationships affect other elements of the new definition of health. People who are socially integrated, by virtue of their relationships, live healthier lifestyles overall: They are less likely to drink alcohol excessively, smoke cigarettes, or be sedentary and are more likely to have medical checkups and take medication as prescribed. People with strong relationships are less likely to become depressed and are less susceptible to the untoward effects of stress.[26] The latest research is now suggesting that our relationships have a direct effect on our biology.

Relationships Change Biology

For social relationships to affect illness and longevity, they at some point have to influence biology. But which biological processes are affected and in what ways? Is it really true that social ties decrease host resistance in a way that increases susceptibility to illness? Despite the ocean of research demonstrating that social support is related to health *outcomes,* its credibility in the medical world as a risk factor rests largely on its effects on biological *mechanisms.* In the last decade dozens of studies have emerged that demonstrate the widespread consequences on biology of having or not having social connections.[27] Following are a few of them:

Relationships predict hypertension. One way that a lower level of social support might affect heart disease is through its effects on blood pressure and hypertension. In a study by University of North Carolina researchers, both instrumental and emotional support were assessed in 2,000 African Americans and whites to determine the effects on blood pressure. High levels of instrumental support were significantly related to lower prevalence of hypertension in African Americans, a group known to have a high susceptibility to this disorder.[28]

Relationships predict atherosclerosis. In the Stockholm Female Coronary Risk Study, 130 women who were hospitalized after a heart attack were given detailed cardiac examinations, which included a coronary angiography, a test used to determine the amount of blockage in the coronary arteries. After accounting statistically for smoking, education, menopausal status, hypertension, cholesterol, and body mass, the data showed that women who were less socially integrated tended to have more blocked arteries and significantly more severe blockages than did women who were more socially integrated. Blocked arteries are serious precursors to heart attacks.[29]

Relationships affect immune-system status. Researchers at Ohio State University have discovered that, compared to noncaregivers, individuals who provide care for family members with Alzheimer's disease report high levels of chronic stress and have impaired immune-system responses. This includes lower amounts of natural killer cells, B cells, and T cells, and smaller immune responses to vaccines. Social support, however, reduces these untoward effects. Although many caregivers show some decline in immune functioning, those caregivers who have higher levels of emotional support have smaller such declines over time than do caregivers with less emotional support.[30]

Several studies have also found that social support can help buttress the immune system even in people who are HIV positive. In one study the health of men infected with HIV was tracked for five years. During the first year those with high and low social and emotional support did not differ in CD4 counts, a key measure of immune status. After four years, however, those high in support showed a decline of 24 percent in CD4 counts, compared to a decline of 46 percent in those low in support. After five years the CD4 counts of the high-support group had declined 37 percent and in the low-support group, 64 percent.[31]

Relationships predict stress-hormone levels. As part of the MacArthur Study of Successful Aging, Dr. Teresa Seeman and associates measured stress hormones in relatively healthy older men and women.[32] After accounting for age, chronic illness, weight, smoking, and medication use, they discovered that, compared to men with lower emotional support, men with higher levels of emotional support had lower levels of cortisol, norepinephrine, and epinephrine—hormones produced by stress.

Relationships predict cardiovascular responses to stress. A fascinating line of research on the biology of social support is the experimental manipulation of social support in laboratory studies. In these studies volunteers are asked to perform challenging mental tasks—such as doing math problems in their heads, making an impromptu speech, or carrying out a debate—while researchers measure their physiological activity. Some of the volunteers perform the task alone, while others receive some form of social support. The type of social support that is given has varied considerably across studies. In some studies social support has been the quiet presence of a friend while the volunteer performs the task. In other studies it has been a stranger who provides encouragement while the volunteer performs the task. Regardless of how social support has been manipulated, the goal has been the same: to determine whether the presence (or absence) of a supportive person changes the magnitude of cardiovascular responses to stress. Most, but not all, studies find that it does. Demonstrating the effects of social support on cardiovascular responses to stress is important, since larger responses of this type may increase the risk of heart disease.[33]

In one such study researchers at Cornell Medical College had men and women college students participate in a debate on a controversial topic, such as abortion, either with or without the presence of social support. In the no-support condition, participants had to defend their positions on abortion while two strangers attacked their views and a third stranger watched passively. In the support condition, the third stranger made supportive comments and gestures while the participants defended their viewpoints. Participants who did not receive support had larger increases in systolic and diastolic blood pressure and in heart rate than those who received support.

It is clear from these types of studies that social relationships can determine our health through a variety of pathways: They affect our health behavior, they affect our emotions, and they affect our biological susceptibility to illness. But we all know that close personal ties are not always positive. Let's take a look at the potential downside of friends and family.

Chapter 7

Mixed Blessings:
The Complexity of
Social Relationships

Despite the positive, health-enhancing, and life-affirming aspects of social connections, they present a paradox. They can serve as our greatest source of joy and the genesis of much of our despair. They are routes to emotional peaks and pathways into a psychic abyss. They are simultaneously stress relieving and stress producing.

In his bestselling book *Love and Survival*, Dr. Dean Ornish provides readers with an intimate glimpse inside relationship conundrums. Of his relationship with his parents he writes,

> Growing up in my family, as in many families, the unspoken yet heard message from my parents was this: "You don't exist as a separate person; you are an extension of us. Therefore, you have a great capacity to cause us joy or pain. If you act right, we will be so proud of you. If you don't, we will suffer. If you really mess up, we will *really* suffer—and if we suffer enough, we will die and leave you all alone. Since you don't exist separate from us, then if we die, you'll die, too. And it will be your fault."

Dr. Ornish goes on to write that upon hearing of his decision to take a year off from medical school to do research, his parents told him, "Isn't it ironic? You want to drop out of medical school to do research on stress and the heart and you're giving us heart attacks!"

According to Dr. Lisa Berkman, "Support isn't always a happy ending, a matter of what you can get, but more about being embedded in the whole social system—in a society, in a community, in relationships—that is important. This can be a real burden—but nobody said that the burden isn't what it's really about. Because that *is* the sense of connectedness. It is an interrelationship that is both giving and taking, which is what love is about, what intimacy is about. It's not just about getting support; it's about knowing that you can count on somebody, and knowing that somebody can count on you."[1] To do that you may sometimes have to make yourself available when it isn't convenient.

Unhealthy Social Relationships

Social ties can be toxic, too. When a relationship reinforces unhealthy lifestyle choices, it's anything but salutary. Our relationships serve as sources of information and cultural norms, role models that form the social context for our behavior. People who are more prone to unhealthy behaviors, such as smoking, illicit drug use, excessive alcohol intake, or overeating, often form social bonds that exacerbate their problems. For example, one of the strongest determinants of the initiation of cigarette smoking by adolescents is whether they have peers who smoke.[2]

Likewise, most of us have had the experience of being offered advice by well-meaning family members or friends when we did not want it or need it, or it was the right advice at the wrong time, or it was given in a way that increased our burden rather than decreased it. Close social bonds can be uplifting but can also open the door to destructive criticism and excessive demands. In their comprehensive analysis of the scientific literature on this topic, Drs. Matt Burg of Yale and Teresa Seeman of UCLA report on studies of cancer patients and bereaved persons who experience the efforts of others as unhelpful and distressing. Occasionally such assistance may induce feelings of guilt and distress over the inability to reciprocate; the person feels he or she should give something back but can't. Cancer patients, in particular, may have feelings of dependency and helplessness when help is given. Research has also found that receiving social support may be associated with increased feelings of depression. On the other hand, some cancer patients become more distressed when the support they expect from friends is not forthcoming.[3]

A recent study from the MacArthur Study of Successful Aging

found that healthy elderly men and women with higher initial levels of instrumental assistance were at higher risk for reporting new disabilities over a two-year period.[4] It was not that the people who were more disabled required more assistance, which is often the case; it was that *high levels of assistance preceded and were a key predictor of their self-report of new disabilities.*

Why would instrumental support of this type be related to greater disabilities over time? Dr. Teresa Seeman believes that there could be both psychological and physiological explanations. She says that "because there is already a societal stereotype that growing older will inevitably be associated with greater disability and reductions in functioning, the receipt of assistance engenders a belief in the older person that he or she does in fact need help and cannot do certain things on [his or her] own, as if age equals infirmity by definition. And remember, we measured *self-reported* disability, so it was based on [people's] beliefs rather than [on] objective abilities. It could have been that greater instrumental support, although well intentioned, undermined their sense of self-efficacy that they could accomplish certain things for themselves. It is almost as if the patients were saying, 'Well, they think I need this help, so I suppose I must.'" Seeman also thinks that there could be a physiological explanation for why support is associated with greater disability. "When older adults get more instrumental assistance, they are not forced to do as many things for themselves, leading to less actual physical effort. If increased instrumental assistance results in less activity on the part of the recipient[s], they may in fact lose some functional abilities because of [the fact that] they are using their muscles less. Research with older adults clearly indicates that if you don't use it, you indeed can lose it," she said.

Losing the Connection: When Marriages Fail

There is perhaps no clearer expression of the complexity of social relationships than the marriage bond. Marriage, as I have noted, confers significant health benefits to both spouses. But by no means is marriage a panacea. It can introduce factors that are quite detrimental to health. In fact, marital conflict—separation, divorce, and widowhood—can have a devastating effect on emotional well-being and ranks among the most stressful life events. When parents divorce, children can be caught in the middle, which can often have lifelong emotional and health consequences.

Dr. Howard Friedman and colleagues at the University of California at Riverside have provided some of the clearest evidence on the long-term effects of divorce on children and on the consequences of one's own divorce.[5] Friedman's team analyzed data from the famous Terman Life-Cycle Study of gifted and talented preteen boys and girls (known as the "termites"). They found that divorce had lasting health effects for the children as they became adults. *Children who were younger than twenty-one when their parents divorced had a 30 percent higher risk of death* than did people whose parents remained married. When these findings are projected across a life span, *men and women whose parents divorced while they were children would live on average four years fewer* than those whose parents remained married.

That in itself is startling, but what happens to the health of a couple when their marriage dissolves? To answer this question Friedman and associates classified the "termites" who had reached age forty or more into four groups:

1. those steadily married
2. those inconsistently married (i.e., married, but not in their first marriage)
3. those never married
4. those separated, widowed, or divorced (although there were very few widows in the group)

Not too surprisingly, the steadily married had the lowest rates of premature death. Interestingly, those who never married had a similar mortality profile *if* they had a social network to compensate. The consequences of divorce were dramatic: *Divorced women were 80 percent more likely than the married women to have died, while divorced men were at a whopping 120 percent increased risk for death com-*

pared to married men. Because women have more outlets for getting and giving support, studies show that the loss of a husband is less deleterious to their emotional and physical well-being.[6]

As striking as those statistics are, what was really surprising is what happened to the inconsistently married. Even though they remarried, both men and women in this group had *a risk of death that was 40 percent higher* than that of their steadily married counterparts. It appeared that the negative consequences of a divorce carried over into, and were not canceled out by, the new marriage. According to the authors, "This finding suggests that it may not be marriage's effect as a buffer against stress that is always important. Rather, there seems to be a detrimental effect of previous divorce that is not eliminated when the individuals remarry."

Of course, the context of divorce is more complicated than simply losing social support. All is not rosy within a marriage during the time leading to separation or divorce. Conflict is prevalent, and negative feelings run high. The stress of marital conflict can induce profound decreases in immune-system activity, which could open the door to disease susceptibility. Researchers at Ohio State University, led by Drs. Janice Kiecolt-Glaser and Ronald Glaser, have been studying the biological effects of marital conflict for nearly twenty years. In their studies married couples are admitted to a hospital research unit for twenty-four hours, where they undergo psychological tests and receive questionnaire assessments of their marital satisfaction and mood. In the laboratory portion of the study, biological measures are taken unobtrusively (e.g., using indwelling catheters to take blood samples continuously) from the couples as they engage in thirty minutes of marital conflict. That is, based on prior interviews and the questionnaire responses of the couples, the researchers ask them to discuss and attempt to resolve two or three marital issues that were deemed the most conflict producing. Their interactions are videotaped and scored using a coding system called the Marital Interaction Coding System (MICS). The MICS measures the extent of hostile and negative behaviors exhibited, which include things such as criticisms, interruptions, put-downs, denying responsibility, making excuses, and disapprovals. One of the most consistent findings is that the more such negative and hostile behaviors exhibited by the couples, the higher their stress hormone levels and the lower the immune-system responses.[7] Whether it is increased stress or lack of emotional support, the point here is that being in a marriage, or in any close interpersonal relationship, can carry health risks, not simply health benefits.

ARE WE BECOMING DISCONNECTED?

As important as social relationships are to our health, it is startling that we are actually becoming *less* connected to one another than ever before. We have increasingly become an isolated, loner society, as evidenced by dramatic drops over the last two or three decades in activities such as civic involvement, volunteerism, philanthropy, and religious participation, and in our perception of others as trustworthy and honest. Our engagement in family activities and social gatherings with friends has also declined precipitously. Our time is increasingly spent on solitary activities rather than interpersonal ones. Consider these statistics from *Bowling Alone* by Harvard professor Dr. Robert Putnam.[8]

- In the last two decades there has been a 45 percent decline in the frequency of friends' entertaining or visiting one another in their homes.
- Between 1981 and 1999 there was a 50 percent decline in the average frequency of card playing among American adults, an intrinsically social activity.
- Involvement in religious activities and church attendance have declined between 25 and 50 percent in the last four decades, virtually erasing the boom in religious participation that immediately followed World War II.
- In the last two decades the percentage of families who report that they usually eat dinner together declined by approximately a third, from 50 percent to 34 percent.
- Participation in civic, volunteer, and work-related organizations (e.g., professional societies, unions) has declined significantly in recent decades.

- Once a standard of American social life, even the bowling league is beginning to vanish. Between 1980 and 1993, participation in bowling leagues plunged more than 40 percent. However, individual bowling increased 10 percent during the same period.
- Between 1960 and 1999 there was a 20 percent decline in the number of people who would acknowledge that "most people can be trusted," with comparable declines in people who would say that "most people are honest."

These and other indices of community and social involvement and perception form the basis of what social scientists call "social capital." As you can see from the statistics above, social capital has been on the decline. Scientists do not know precisely why, but believe it is attributable to a combination of factors, such as people spending time on the Internet, an increase in television viewing, economic pressures that require more time at work and more dual-career households, and less social involvement among younger generations. The decline in social connectedness is more than interesting sociology—it carries larger significance for the functioning of our society. For one thing, communities that are higher in social capital tend to be better places to live for all of their citizens. As an example, they have more opportunities and show better outcomes for children on a number of measures of their well-being, have safer and more productive neighborhoods, experience improved economic prosperity, and engender greater emotional well-being for those who live there. More important, communities that are higher in social capital are also healthier ones.[9]

Joining, Trusting, and Longevity

Dr. Ichiro Kawachi of Harvard's School of Public Health is the leading authority on the health aspects of social capital. In two innovative studies he and his colleagues have examined whether statewide measures of civic involvement and trust were related to physical health and longevity. The studies addressed a key question related to social connectedness: Do residents of states that are higher in "sociability" and trust live longer than do residents of other states? The researchers analyzed data from thirty states that were part of the massive General Social Surveys (GSS), conducted by the National Opinions Research Center, and the Behavioral Risk Factor Surveillance System (BRFSS), conducted by the Centers for Disease Control and Prevention. To address the sociability issue, Kawachi used the GSS data to rank each state on the degree to which its residents, on a per capita basis, joined civic organizations, including sport and church groups, labor unions, fraternities and societies, and hobby groups, among others. Community involvement ranged from a low in Arkansas of 1.2 groups per capita to a high in North Dakota of 3.5 groups per capita. The researchers found that in 1990 community involvement was inversely related to mortality. The higher the per capita group membership, the lower the statewide mortality from all causes combined and deaths from heart disease, cancer, and infant mortality when examined separately. Amazingly, every one-unit increase in per capita group membership was associated with 66 fewer deaths for every 100,000 people.[10]

Similarly, using data from the BRFSS, which involved over 167,000 participants, the researchers found that individuals in states lower in social capital were significantly more likely to rate their health as "poor or fair" than were residents of states higher in social capital.[11]

In the GSS study, Kawachi and colleagues also sought to determine if statewide trust was predictive of death. To assess trust, participants were asked whether they believed statements like these were true: "Most people can be trusted," "You can't be too careful in dealing with people," "Most of the time people try to be helpful," and "Most people would try to take advantage of you if they got the chance." The level of social trust was strongly predictive of mortality. As shown in Figure 7, states that had higher percentages of people believing that "most people would try to take advantage of you if they got the chance" were also states with the highest mortality. As with community involvement, higher levels of social trust were related to fewer deaths from all causes combined and to single causes of death such as heart disease, cancer, stroke, homicide, and infant mortality.[12]

Kawachi believes that things like social capital matters for health because it might be a community-level analog to concepts like social support and social connectedness. That is, *communities* can be characterized as having denser networks of social connection, or higher levels of cohesion and solidarity, just as *individuals* can be characterized as socially isolated or well connected. Although parallels can be drawn between individual-level social support and community-level social capital, Kawachi believes that social capital has a separate and independent effect on people's health—over and above their own individual levels of sociability. Kawachi states that "an individual might be socially isolated, but nonetheless benefit from the level of social connectedness between her neighbors, because their neighbors watch over each other's homes to prevent crime, or because neighbors might lend a hand in clearing snowy sidewalks, or because neighbors might organize to improve the quality of schools, prevent junk food shops from moving into the neighborhood, or lobby against the closure of a local fire station."

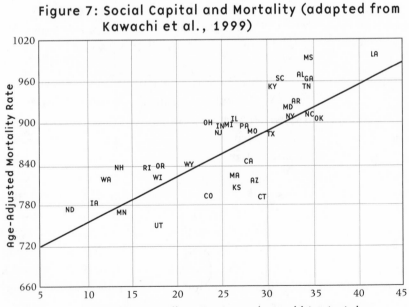

Figure 7: Social Capital and Mortality (adapted from Kawachi et al., 1999)

Percent Responding: Most people would try to take advantage of you if they got the chance.

CONCLUSION TO PART III

When it comes to health, relationships indeed matter. However, that statement of scientific fact does not tell us what we should *do*, what courses of action we all should take to improve our health or prevent disease. I came face-to-face with this dilemma recently when working on a public-service announcement for a proposed television campaign about social relationships and health. My colleagues and I wanted to communicate to a wide audience the health-enhancing effects of social ties, but we struggled with the breadth of our message: What specific forms of social ties should we advocate for? Should we tell people to make new friends, get a confidant, become less socially isolated or help others become less isolated, get married, stay married, spend more time with family, or give and be open to receiving more assistance from others? We debated each of the scientifically valid options but realized that each did not apply to every individual all the time. For example, what about people who are socially isolated because of illness or disability? Telling them to become less socially isolated for their health might place an undue burden on them. Married people on average have better health and live longer than single people, but bad marriages with lots of conflict are biologically toxic—but then again, divorce can have negative health consequences as well. Having a large social network is generally good from a health perspective, but being embedded in a large network can also be burdensome and the source of stress. What to do?

When it comes to social relationships, it might be best to recognize their well-established scientific benefits while at the same time falling back on that old 1960s saying: "If it feels good, do it." By this I am suggesting not that we devolve into hedonism but rather that we should be sensitive to the intrinsic value and emotionally rewarding nature of relationships, beyond their effects on health. Quite simply, there is a common thread of experience that occurs when we are with others whom we care about, whom we have a desire to help, who are a help to us, or with whom we wish to share experiences. That common thread of experience is feelings of connectedness, closeness, love, altruism, security, and a sense of belonging. The by-product of these feelings is likely to enhance health and longevity.

We have much more to learn scientifically about social relationships and health. The really big question seems to be this: If people who are low in social relationships are taught ways to increase them, will their health be better over the long term? It would seem the obvious answer is yes, and there is some indication from research with cancer patients that support-group interventions may indeed increase longevity.[13] But the research on the long-term health effects of social-support interventions in large, randomized national studies is just evolving.

The next part of this book describes another determinant of health that falls under the broader rubric of "social factors." Specifically, it addresses one of the oldest and most puzzling findings in epidemiology: that a person's position on the economic ladder is among the strongest determinants of his or her health.

PART IV

PERSONAL ACHIEVEMENT AND EQUALITY: LEARNING, EARNING, AND SURVIVING

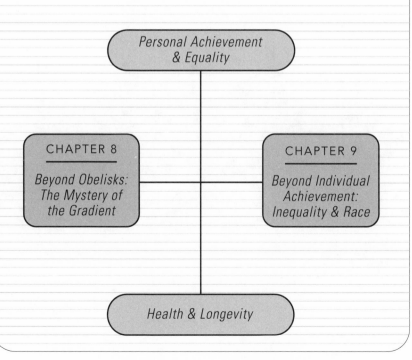

Personal Achievement & Equality

CHAPTER 8

Beyond Obelisks: The Mystery of the Gradient

CHAPTER 9

Beyond Individual Achievement: Inequality & Race

Health & Longevity

Learning is . . . a refuge in adversity, and a provision in old age.

—Aristotle

Our inequality materializes our upper class, vulgarizes our middle class, brutalizes our lower class.

—Matthew Arnold

This is a photograph of an ancient cemetery located outside the Cathedral Necropolis in Scotland. It was the site of an unusual and creative study, one that highlights in stark terms one of the most powerful determinants of our health today—affluence.

The study was conducted in the early 1990s, by the Scottish researcher Dr. George Davey Smith, who used the cemetery to answer a simple but strange question: Could the height of the obelisk-shaped grave markers reveal anything about the longevity of the deceased? To answer this question he compared the height of each grave marker to the age of the person buried beneath when he or she died. What he found was a striking correlation between the heights of the obelisks and the ages at death. The taller the marker, the longer the life. Men and women commemorated by the tallest grave markers had lived significantly longer than had those with shorter markers. Men had lived more than three years longer and women about seven years longer.

Does the height of Victorian grave markers actually have anything to do with longevity, or is this one of those serendipitous correlations that really means absolutely nothing? Obviously there is no cause/effect connection between grave markers and death. The height of the grave markers did not cause the deaths. The researchers were interested not in grave markers per se but in a more profound phenomenon. They were looking for a historical corollary, an ancient harbinger of one of the most consistent, pervasive, and least understood findings in contemporary health science: ***that affluence (or a lack thereof) affects health and longevity.*** From historical documents

Davey Smith learned that there was quite a pecking order when it came to the obelisks. The markers were a sign of family prestige and wealth, with taller markers symbolizing more of both. The prices of the obelisks rose exponentially with their height, so only the more affluent families could invest in the more commanding markers. The hypothesis that family wealth increased longevity was confirmed by what was found in that cemetery: On average, affluent people live longer. They did then, and they do now.[1]

A Healthy Position

The next two chapters are different from the others in this book because they have less to do with you and your personal relationships and more to do with your position in the economic universe compared to that of others. In America, and in many other developed countries, great value is placed on egalitarianism and equality. We like to think that we are all equally valued and worthy of everything life has to offer. Yet there is a contradiction of sorts. Our egalitarian principles coexist within a society that is fundamentally hierarchical. We order and rank ourselves based on our income, our educational attainment, and even the status of the jobs we hold—conditions that make up what scientists call *socioeconomic position* (SEP). Your position in the economic universe establishes the boundaries for decisions large and small, such as where you shop for clothes, where you live, and where your children will go to school. Your SEP is about where you stand on the socioeconomic ladder relative to others. In Chapter 8 you will see that your position in the socioeconomic hierarchy—whether high or low, comfortably middle class, or on the cusp of affluence—affects your health.

The reasons that economic factors influence health are far from indisputable. There is much debate among scientists about how SEP "gets under your skin" to affect your longevity. Is it that wealthier people have better health care, less stress, better diets? Do they get more exercise and live in safer neighborhoods? Of course they do. But these "usual suspects," as I call them, explain only part of the story. There is more to the SEP effect than health care, lifestyle, or living conditions. Exactly what, is not completely clear. In spite of not knowing why SEP has such powerful effects on health, we do have options for minimizing its untoward effects.

This section of the book is also about another contradiction to our egalitarian principles: the health impact of the widening gap between those at the extremes of wealth and poverty, those at the very top of the economic ladder and those at the bottom. If you are at neither extreme, as is the case for most of us, you may think that the distance between these two diametrically opposed groups does not relate to you at all. However, as I discuss in Chapter 9, when the gap between the rich and the poor is large, the health of everyone suffers, including that of the great majority of us in the middle.

Previously I have used stories of real people to illustrate and put a personal face on the concepts being described. Now I depart from that practice in favor of using statistics from representative surveys of the American public to provide you with a national, rather than individual, snapshot of our overall economic well-being.[2] I call these snapshots "Life Along the Economic Ladder." The first one follows:

LIFE ALONG THE ECONOMIC LADDER

17 percent—those of us who've had times when we did not know where our own or our family's next meal was coming from

25 percent—those of us who've had to cut down on some important food items, such as cereal, fresh fruit, or vegetables, because of costs

34 percent—those of us who've had to work extra hours or an extra job in order to make ends meet

40 percent—those of us who've sometimes found rent, mortgage, or utilities difficult to pay because of the pressures of other needs

56 percent—those of us who've had to cut down on vacations or evenings out because of other needs

71 percent—those of us who've believed that it's very hard, or fairly hard, for the average person to provide for themselves or their families.

adapted from Miringoff et al. (2001).

Chapter 8

Beyond Obelisks:
The Mystery of the Gradient

So how is SEP determined, and what is the modern-day proof that it's important to your health? Most scientists measure SEP with one of three measures: (1) educational attainment, usually assessed by years of formal education, (2) individual or family income, or (3) occupational status or prestige. Studying SEP this way has a long tradition in the social sciences, but it is experiencing a renaissance, following the discovery that the connection between SEP and longevity is more pervasive and complex than ever imagined. Not only is there a line of demarcation between the rich and the poor, but there is also a sliding scale of health status that extends from the richest and most educated at the top to the poorest and least educated at the bottom. With every step down the economic ladder, longevity is compromised and health problems increase. Believe it or not, health disparities exist, at least statistically, even between the supremely rich Bill Gateses of the world and the merely ultrarich Donald Trump types. This phenomenon is called the SEP/health gradient. *Webster's* defines a gradient as ascending or descending with a uniform slope.[1] In this case it means that the higher you are on the SEP ladder, the better your chances for health and longevity. Figure 8 shows a representation of the SEP/health gradient. The gradient is applicable for nearly every cause of death, every illness, and every disabling condition.

The gradient does not discriminate: It affects men and women, people from every racial and ethnic group, and anyone who lives in an industrial nation.[2] In fact, SEP is perhaps the industrial world's strongest and most ubiquitous risk factor for illness and death. Nothing affects more diseases or touches as many of us as SEP does. The

Figure 8: Representation of the SEP/Health Gradient

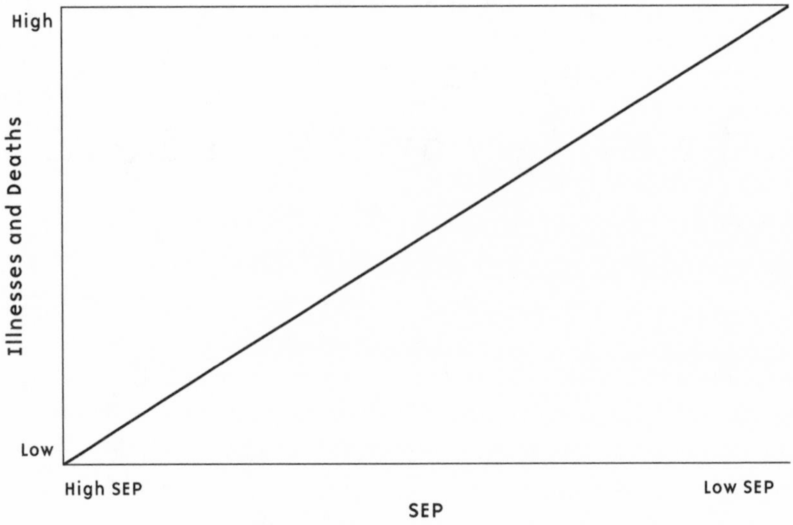

health effects of SEP are independent of other risk factors, such as smoking, race, age, and gender.[3] In addition, among persons who already suffer from illnesses such as heart disease, SEP is a predictor of who survives.[4] The study of the SEP health gradient is one of the hottest and most challenging areas of the new health science. Because SEP touches so many dimensions of our health, including relationships, emotions, behavior, and biology, it has attracted scientists from many disciplines, who are now working together to reveal its mysteries. Here are some specific reasons SEP is now taken so seriously by so many researchers:

SEP predicts mortality and morbidity. Of all of the studies of SEP and health, the Whitehall study is perhaps the most famous. It began in the late 1960s as a comprehensive exploration of the origins of health problems in more than 18,000 British male civil servants (government workers).[5] Led by Sir Michael Marmot, the Whitehall study was ideal for examining the relationship between SEP and health. First, it had a large number of healthy participants, who could be tracked over the course of many years to assess illnesses and deaths. Second, since the participants were civil servants, detailed information about their occupational "rank" was available. These rankings—or "grades," as they are known in government vernacular—ranged from top administrators to unskilled manual workers. Third, in addition to illnesses themselves, the study assessed risk factors for illnesses, such as smoking, high blood pressure, excessive body weight, high cholesterol, and elevated hormone levels.

Figure 9 shows death rates after ten years for the participants in the Whitehall study in three categories: (1) from all causes, (2) from coronary heart disease (CHD) alone, and (3) from illnesses other than CHD.[6] Participants in the lower-ranking jobs, especially support staff (e.g., messengers and other unskilled workers) and clerical workers, had a substantially higher death rate during the ten-year evaluation than did those in the higher-ranking professional, executive, and administrative positions. This SEP mortality gradient remained strong even at the twenty-five-year follow-up.[7]

Figure 9: Socioeconomic Position and Mortality in British Civil Servants (adapted from Marmot et al.,1984, Rose and Marmot, 1981, and Kaplan and Keil, 1993)

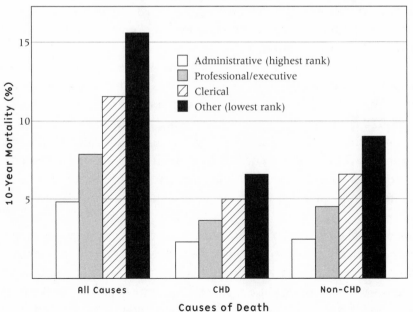

Overall, people in the lowest-grade jobs had a *212 percent higher death rate* than did those in the highest-grade jobs. One reason the Whitehall study findings are so impressive is that none of the participants were in abject poverty, and all of them had access to medical care through the British National Health Service. Consequently the gradient could not be explained simply on the basis of access to health care, which is a typical explanation. Although more affluent workers could have supplemented their medical coverage from private sources and may have received a higher quality of care, there was no evidence that illnesses were undertreated at the low end of SEP.[8] In fact, other research has demonstrated that SEP is equally predictive of illnesses that are amenable to medical care and those that are not.[9]

Another study to show the SEP mortality gradient is the U.S. Longitudinal Mortality Study that followed more than 1 million people for six years, examining the connection between education, income, and mortality. Educational attainment was found to be a strong determinant of mortality. For example, black and white men and women who had not completed high school were between *44 and 78 percent more likely to die* (depending on which groups were compared) during those six years than were those with five or more years of college.[10] Similar trends occurred for income levels: The higher a participant's income, the lower the likelihood of death during the study. This held true for blacks and whites.

The U.S. Longitudinal Mortality and Whitehall studies are just two of many that have shown a significant association between SEP and mortality.[11] Given its pervasive effects on mortality, it is not surprising that SEP is a powerful determinant of specific illnesses. Table 1 shows a few of the health conditions that become more prevalent with lower positions on the SEP ladder. The gradient is found for all types of chronic diseases, including cardiovascular, gastrointestinal, musculoskeletal, neoplastic, psychiatric, pulmonary, and renal diseases.[12]

SEP predicts survival in people undergoing medical treatment. Once a health problem is diagnosed and treatment begins, it seems theoretically that SEP would become irrelevant. Medical treatment should be the great equalizer, benefiting everyone in the same way, but several studies have shown that response to medical treat-

Table 1.
EXAMPLES OF THE HEALTH PROBLEMS ASSOCIATED WITH SEP

Heart disease	Arthritis	Back ailments
Gallbladder/ liver disease	Epilepsy	Cancer
	Unintentional injuries	Sudden infant death syndrome
Stroke	Mental illnesses	Mental retardation
Diabetes	Chronic bronchitis	Substance abuse
Low birth weight	Hearing impairments	AIDS progression and survival
Emphysema	Stomach ulcers	Blindness
Infant mortality	Kidney disease	Hypertension

ment is not exempt from the influence of SEP. Redford Williams and colleagues at Duke Medical Center studied more than 1,300 patients who were undergoing treatment for severe blockage of at least one artery.[13] It was found that SEP, measured here by income, was strongly related to survival. People with higher incomes survived longer. Over the course of five years, *it was discovered that patients with annual household incomes below $10,000 were nearly twice as likely to die as those with incomes of $40,000 or more.*

SEP affects biological processes. For SEP to have such profound and pervasive effects on health and survival, it stands to reason that it must somehow affect the body's biological systems. So in recent years a great deal of effort has been devoted to understanding how SEP affects biology. As expected, scientists have confirmed the interconnection between SEP and biology that could ultimately lead to a variety of illnesses: The lower the SEP, the higher the biological risk profile. Lower SEP is related to biological changes that increase risk for a host of illnesses, including heart disease, diabetes, and others listed in Table 1. For example, lower SEP is associated with:

- higher blood pressure
- higher levels of bad cholesterol (low-density lipoproteins)

- higher serum triglycerides
- impaired glucose tolerance
- greater insulin resistance
- greater blockage of coronary arteries
- higher plasma fibrinogen
- augmented cardiovascular stress responses
- higher allostatic load scores
- more rapid progression of HIV infection[14]

For example, two studies, one from the United States and one from the United Kingdom, illustrate the impact of SEP on biology, even before diagnosable diseases are present. Drs. Linda Gallo and Karen Matthews and their colleagues at the University of Pittsburgh conducted an analysis of data from the Healthy Women Study, a prospective study of the biological and behavioral risk-factor changes associated with menopause.[15] The Pittsburgh team wanted to know if educational attainment was associated with the degree of calcification in the coronary arteries and the aorta. Calcification is one of several biological measures of the degree to which a person's blood vessels are blocked due to atherosclerosis. Measuring calcification permits an assessment of early-stage heart disease. Educational attainment was strongly associated with aortic calcification and moderately related to coronary calcification. The findings for aortic calcification are shown in Figure 10.

Severe aortic calcification occurred in twice as many people with lower educational attainment compared to those higher in education. Less than 15 percent of the people with college degrees were found to have severe aortic calcification; however, *calcification occurred in over 30 percent of those with a high-school education or less.* The association of SEP with calcification could be a consequence of SEP's effects on many of the biological factors listed in Table 1, as well as other factors.

In London, researchers Drs. Eric Brunner, Michael Marmot, and colleagues examined data from the Whitehall study to investigate a possible link between SEP, as defined by employment grade, and metabolic syndrome.[16] Metabolic syndrome is a clustering of biological risk factors for heart disease and involves at least three or more of the fol-

Figure 10: SEP and Severe Aortic Calcification
(adapted from Gallo et al., 2001)

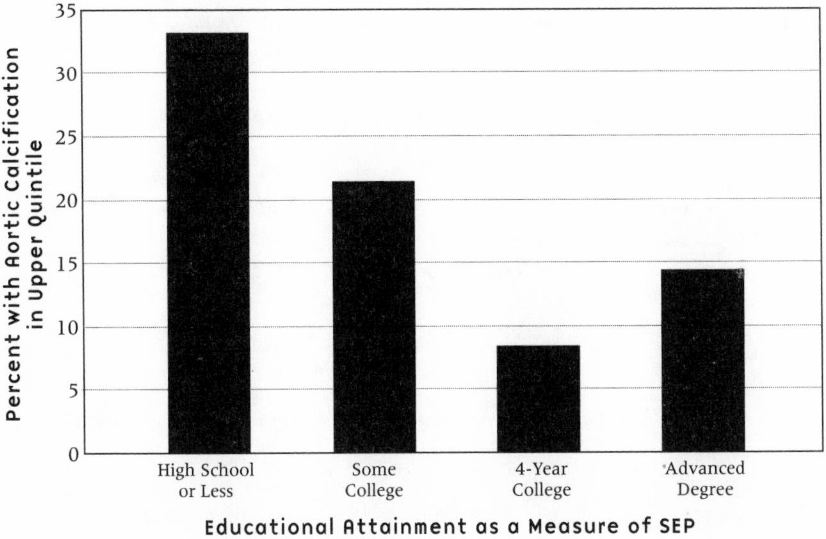

Educational Attainment as a Measure of SEP

lowing: impaired glucose tolerance, high systolic blood pressure, high fasting triglycerides, low HDL (the good) cholesterol, or a high waist-to-hip ratio (indicating high central obesity). *The odds of having metabolic syndrome were over twice as high for people at the lowest employment grade,* compared to those at the highest. This was true for both men and women.

HOW DOES SEP GET UNDER YOUR SKIN?

Although scientists have learned much about the gradient, the $100,000 question remains: Why does it matter to your health how far you progressed in school, how much money you make, or the status of your job? How does your SEP get under your skin to affect your longevity? In this section I explore some of the intriguing hypotheses about the whys and hows of the gradient.

LIFE ALONG THE ECONOMIC LADDER

31 percent—*those of us who at times worry that we will not be able to cover health-care costs if we or a family member becomes ill*

> *16 percent—those of us earning more than $50,000 a year who feel this way*

> *40 percent—those of us earning less than $50,000 a year who feel this way*

31 percent—*those of us who worry about paying for the costs of medications*

> *19 percent—those of us earning more than $50,000 a year who have such worries*

> *39 percent—those of us earning less than $50,000 a year who have such worries*

52 percent—*those of us who sometimes feel rushed during medical visits, as if the doctor is watching the clock*

> *43 percent—those of us who are college graduates who feel this way*

> *56 percent—those of us with a high-school education or less who feel this way*

adapted from Miringoff et al. (2001).

The Usual Suspects

One of the first things scientists do when trying to understand a new health risk factor is to determine if it is related to any of the established risk factors. The logic here is that the new risk factor might be affecting health by impacting one of the risk factors we already understand. These known risk factors include smoking, physical inactivity, lack of access to health care, excessive alcohol intake, and an unhealthy diet (e.g., low intake of fruits and vegetables, high fat and salt intake). People lower on the SEP ladder, on average, tend to have more of these risk factors than do those who are higher.[17] Table 2 shows findings from the Whitehall study comparing health behaviors in the highest and lowest employment grades. Notice that both men and women in the lowest

Table 2.

HEALTH BEHAVIORS BY EMPLOYMENT GRADE
IN THE WHITEHALL STUDY

Health Behaviors	Lowest Grade		Highest Grade	
	Men	Women	Men	Women
Percentage current smokers	33	27	8	18
Percentage reporting no exercise	30	31	5	12
Percentage using skim or semi-skim milk	21	34	44	39
Percentage using mainly whole-meal bread	32	35	47	57
Percentage not eating fruit or vegetables daily	61	43	34	17

grades had significantly higher rates of smoking, lower rates of exercise, less use of skim milk, less use of whole-meal (or whole-grain) bread, and less consumption of fruits and vegetables.[18] It may occur to you that most of the risk factors reflect behavioral "choices" that we make. No one tells us we have to smoke, eat unhealthy foods, or be sedentary. But why would unhealthy choices be made more often by those at the low end of SEP? One reason might be that they are less knowledgeable about lifestyle factors and health than those higher in SEP. But this could also demonstrate the incredible power of the social context in shaping our choices. For example, people of lower SEP may have less access to healthy foods, either because they are not stocked by neighborhood stores or because they are cost-prohibitive.[19] Likewise, cigarette and alcohol advertising is more prevalent and exercise facilities less so in lower-SEP areas.[20] So the greater number of risk factors and higher rates of death and illness in this group are not surprising.

It stands to reason that if those of us in the health professions could just somehow help people at the low end of SEP have the same risk profile as people at the top of SEP, everyone's health would be equal and the gradient would disappear. Unfortunately, it would not. True, an equaling out of risk factors such as smoking and physical inactivity across groups would certainly make the gradient less steep, and this is a meaningful goal, but the linear relationship between SEP and health would still exist. Study after study has shown that even when all the traditional risk factors are taken into consideration, the gradient is lessened, yet it remains potent.

Some scholars have argued that it is not so much your SEP that puts you at risk relative to others, it is the income you lose when you get sick. That is, illness leads to some "downward drift" of your SEP. So part of the reason for the link between SEP and illness is that illness leads to a loss of SEP, rather than to a lower SEP's causing illness. And if you have a chronic illness, it may be harder for you to move up the SEP ladder. Sure enough, rigorous analyses have shown that chronic illness leads to reduced earnings and savings for many and accounts for some of the income/health gradient.[21] This is especially true for families that are already economically taxed and without health insurance; a serious illness can have devastating financial consequences. Schizophrenia is a case in point—the onset of symptoms occasions a deteriorating economic standing, rather than the other way around. However, downward drift does not explain most of the effects of SEP on health. In many studies the income or education of healthy people predicts their health years down the road. So in the causal chain of events, income and education level came before the illness. Take education as a specific example. We typically complete our formal education when we are younger and healthier. That level of education subsequently predicts our future health and of course education does not go down when we become sick. Thus, while downward drift is a reality, it does not explain all aspects of the gradient.

Thus the mystery of the gradient continues. To solve it, scientists are expanding the search for clues and developing creative hypotheses about what might be the truly toxic features of the gradient.

Economic Security as Stress Management

Without a doubt, stress is a prominent feature of nearly every theory of how SEP exerts its effects on health. Not long ago I got a personal glimpse of why. I was scheduled to give the keynote address at a conference in front of about eight hundred people. It was a big presentation, so I'd put a lot of effort into preparation. I develop my presentation

---

LIFE ALONG THE ECONOMIC LADDER
---

22 percent—*those of us who do not go out at night because of concerns about our safety*

> *26 percent—those of us earning less than $50,000 a year who have such concerns*

> *16 percent—those of us earning more than $50,000 a year who have such concerns*

45 percent—*those of us who worry about the impact of money difficulties on our health*

> *58 percent—those of us earning less than $35,000 a year who have such worries*

> *31 percent—those of us earning over $75,000 a year who have such worries*

57 percent—*those of us who feel stress a great deal of the time*

> *62 percent—those of us earning less than $50,000 a year who feel this way*

> *56 percent—those of us earning more than $50,000 a year who feel this way*

88 percent—*those of us who have friends with whom we can discuss a serious personal or family problem*

> *84 percent—those of us who have a high-school education or less who have such friends*

> *93 percent—those of us with a college education who have such friends*

adapted from Miringoff et al. (2001).

---

slides on the computer. Then I send the file to my assistant, who has the slides made. This time I was a little late finishing the presentation, and as a result the turnaround deadlines were tight. I was expecting a Saturday home delivery of the slides so I would have them for my Sunday flight to New York and my Monday-morning presentation.

By Saturday at 4:30 P.M. I realized that the slides had not arrived. I hadn't thought about them until then, because my assistant is very reliable and had sent packages to my home before. But clearly something was wrong this time. Compounding the problem was that I did not have my assistant's home telephone number, a tracking number, or even the name of the specific overnight-delivery service that was used. I called them all, but no package for me could be found.

For some speakers this situation would be no big deal. They would simply use transparencies, give a computer-based PowerPoint presentation, or speak from notes. The nature of my presentation precluded any of these options. I couldn't even speak from notes—because there were no notes! I speak exclusively and extemporaneously from the prompts provided by my slides, some of which are magazine covers or photographs and hard to duplicate. I'm the first to admit that I *am* a little too obsessive when it comes to my presentations. So on that Saturday afternoon I just could not accept the idea that I wouldn't give the kind of presentation I'd worked so hard to prepare. The stage was set here for me to have a major stress reaction (see Part VI for details on stress reactions). As I thought more about my predicament and saw my options diminishing by the minute, my heart rate started heading north, and I'm certain my blood pressure and hormones were not far behind. My emotions were along for the ride: I was experiencing the full spectrum of anxiety, frustration, and the beginnings of a depressed mood.

My unsuspecting wife, having just awakened from a cold-induced nap, came into my office at about 5:00 P.M. and found me in a slight panic. Together we set out to find someone who could make slides Saturday night and deliver them Sunday. Via the Internet Elizabeth found someone in Cleveland who could not only make the slides but drive them to the Cleveland airport and put them on a plane to New York, where I could pick them up after my plane landed.

His services, of course, were not free. In fact, getting me out of this jam was going to cost me a lot—just enough to hurt. But I was willing to do it. The point, however, is not that I was simply *willing* to pay my own money to solve this problem, which in the grand scheme of unfortunate life experiences was not at all catastrophic. The point is that I was fortunate to have a credit card that made it possible for me to do it—and to short-circuit my stress reaction in the process.

But what if things had been different? What if I did *not* have the resources to pull out of this admittedly relatively minor jam? I would have survived and improvised a presentation. But the emotional fall-out would have lingered until my talk was over. Having the ability to cover the cost provided me with an expedient emotional way out. It helped me cut off the cascade of events leading to the negative emotions and their biological consequences.

When we think about wealth, we tend to envision multimillionaires. But wealth can be defined simply as the amount of money you have at your immediate disposal. Wealth can be one dollar or a billion dollars. Wealth determines what kind of economic safety net we have. The more wealth, the tighter the weave of that safety net. With little wealth the weave of the safety net is loose and less secure, and a crisis may cause you to fall through.

Naturally, when you move down the economic ladder, money worries and financial crises increase. Currently, while more than 50 percent of people in the United States who earn less than $50,000 a year find it difficult at some time to pay rent, mortgage, or utilities because of the pressure of other needs, only 23 percent of Americans making more than $50,000 a year feel this way.[22] Not surprisingly, the lower you are in the SEP hierarchy, the less likely you are to be able to "buy" your way out of life's unpredictable crises, crises much more significant than my slide episode. Like when a child is sick and a parent has to decide whether to stay home and risk losing a job because there's no vacation or sick leave. Like when the heating bill goes up and there's no extra money. Like when a teenager wants to go to an expensive private college, but a state school is more congruent with the budget. This is not to suggest that chronic stressors are few for those higher up on the SEP scale, but the financial ability to cope with them is better.[23] There are, of course, many methods for coping with the stresses and strains of life that have nothing to do with economics. During my slide episode I could have used any number of stress-management techniques to lower my anxiety. But the point is that because of limited economic resources, people at the lower end of SEP are more frequently confronted with stressful circumstances that simply don't affect others. Over time this has to take a toll.

And the Winner Is . . . : The Social Order of Things

In 2001 researchers from Toronto published a thought-provoking study on the life expectancy of actors and actresses who had been nominated for an Academy Award.[24] Three groups of actors were compared: (1) those who had won an Academy Award, (2) those who were nominated but did not win, and (3) nonnominated cast members, of the same sex and age, who were in the same films as the nominees in groups one and two. Amazingly, it was discovered that Academy Award winners lived significantly longer than did performers who were nominated but did not win, as well as those who were never nominated. Comparing winners to nonnominated performers, *Academy Award winners lived an average of four years longer.*[25] Winning the award multiple times was associated with even greater longevity. To put these findings in perspective, we would have to cure and eradicate all forms of cancer across the globe to achieve a similar overall gain in life expectancy.[26]

Why on earth would winning an Academy Award confer longevity advantages? It is difficult to say for sure, but the reason may involve *social ordering,* one of the hottest theories about SEP. Essentially, social ordering is the notion that something is inherently unhealthy about being lower on the social and economic totem pole than your personal reference group—those people with whom you often compare yourself as you chart your progress in life. Dr. Nancy Adler of the University of California at San Francisco explains that social ordering "is the relative status of people in a group or population on any salient dimension. Unlike non-human animals where there is a single dominance hierarchy, humans compare themselves to others on a variety of dimensions and activities. Indices of SEP like income and education are ways that we use to place ourselves and others on a social hierarchy. Where we view ourselves on those or other hierarchies may be what matters most to our health."[27]

Social ordering is an innovative concept because it is common to each of the main measures of SEP—income, education, and occupational status are all ways to ascertain where we stand relative to others. Therefore, social ordering is really about our perceived social standing relative to those with whom we compare ourselves.[28]

Social ordering is a provocative concept also because it provides a single explanation for why people who are ostensibly well-off are likely, on average, to have shorter lives and more health problems than those who are more affluent and why college graduates tend to live longer than high-school graduates. The higher people are on the income or education ladder, the more opportunities they have for favorable comparisons to others along certain dimensions. Even professional achievements and accomplishments can be looked at as another way of ordering people. The Academy Award study may be an indication of this. Being nominated for an Academy Award is unquestionably a tremendous honor, and it places the nominee among the elite in the movie business. On the one hand, then, the nomination places the nominee higher in the Hollywood pecking order than those not nominated. At the same time, being nominated but losing means that suddenly you are on the low end of another pecking order. If your principal comparison group is the winners of the Academy Award, then your mere nomination falls woefully short.

I got a taste of a pecking order just after I completed my Ph.D. training. It felt great to have finished graduate school and make it to the top of the educational ladder. My friends, family, and members of my parents' church expressed their sincere congratulations and pride in my accomplishment. But that great feeling soon faded as I began working in academia, first as a research fellow and then as a young professor. In this world having a Ph.D. was no big deal, since *everybody* had a Ph.D. or some other doctoral degree. What's more, as a newly minted Ph.D., I was at the low end of the Ph.D. pecking order. I actually remember feeling a little down one day at the thought that after all those years of graduate school, I had made it to the top of one ladder only to be at the bottom of another. Whether I felt elated that I had obtained the Ph.D. or deflated that I was on the low end of the Ph.D. ladder depended entirely on whether I compared myself to the population at large or to my senior colleagues. In many respects SEP may have less to do with your *actual* position in some hierarchy and more to do with your *perception* of your position.

Studies with humans and animals indicate that social standing is a powerful determinant of health.[29] So-called white-coat hypertension is a case in point. About 20 percent of patients whose blood pressure is elevated when a physician measures it in a clinic have normal measurements outside the clinic. Theoretically, the elevations are caused by the clinical environment or by having the blood pressure measured by an individual perceived as having high status and power, which could be anxiety producing.[30, 31]

More recent evidence about the importance of social ordering comes from the research on subjective social status. Two studies by Nancy Adler's research team in San Francisco indicate that where we rank ourselves relative to others may be even more important than our education or income. To determine how study participants viewed themselves, researchers showed them a drawing of a ladder with ten rungs and asked them to do the following:

> Think of this ladder as representing where people stand in the United States. At the top of the ladder are the people who are the best off— those who have the most money, the most education, and the most respected jobs. At the bottom are the people who are the worst off— who have the least money, least education, and the least respected jobs or no job. The higher up you are on this ladder, the closer you are to the people at the very top and the lower you are, the closer you are to the people at the very bottom. Where would you place yourself on this ladder?[32]

Participants then placed an X on the rung to indicate their status relative to others in the United States, and this was called their "subjective social status."

In one study, which involved 157 ostensibly healthy white women, lower subjective social status was related to self-ratings of poorer health, negative emotions such as depression and anxiety, and to more chronic stress and pessimism. Lower subjective social status was also associated with increased heart rates. Remarkably, lower subjective social status was also more strongly associated with poorer self-rated physical and psychological health than was lower *objective* social status, as measured by a combination of education, income, and occupation. The effects of subjective social status occurred above and beyond the effects of objective social status.[33] In essence, this study suggests that where we place ourselves on the social-status ladder may affect our emotional and physical health even more so than our objective SEP.

In the other study Adler's team explored the health impact of subjective social status in an ethnically diverse group of pregnant women.[34] Again subjective social status predicted self-rated health, but there were some ethnic-group differences. Among white and Chinese-American women, subjective status was a more potent predictor of health than income was. Among African-American and Latino women, income was a more important determinant of health than subjective status was. Why the difference? According to the authors, the Latino and African-American women had much lower incomes than the white and Chinese-American women. They speculated that income may be more important for health than subjective social standing is at the lower end of the income spectrum. This could be because achieving an adequate income is the critical goal and the most meaningful determinant of quality of life for people who struggle financially. But as you move up in income, and your economic viability is more secure, the perception of your relative social standing becomes more salient. Thus social ordering might operate differently for those with varying income levels.

Adler notes, "Subjective SEP allows individuals to determine for themselves what aspects of their lives are central to their social standing. For one person education may be central to [his or her] sense of social standing, while for another [the] job may be more important." This might help her explain why some people who are high in education or income die much earlier than would be expected based on their objective SEP, while many people lower in objective SEP live extremely long lives. A lot of it depends on the criteria one uses to determine one's social standing. Aspects of some people's lives that have little or nothing to do with the worlds of work, income, or education may be crucial to how they view themselves. The man with a low level of education who works in a low-status job might nevertheless view himself as being high in overall social standing because he is a deacon in his church. In many communities being a deacon accords great respect and standing. Would such a person have a statistical likelihood of living longer than one whose subjective social status was based on a similar (and relatively lower-status) educational and work background? Adler notes that it is too soon to make such predictions, but one thing is certain—the ideas of social ordering and subjective social standing have expanded the search for the causes of the SEP/health gradient.

A Matter of Choice and Control

As I have noted, SEP plays a huge, perhaps dominant, role in the options available to us and the decisions we make. SEP, especially income, places the chief constraint on freedom of choice—all options and choices are at our disposal if we can afford them. This constraining and expanding nature of SEP on our choices was pointedly illustrated to me during my brief career as a college basketball player. I developed a close friendship with one of my teammates, Ron (a pseudonym). He and I shared a range of interests, from studying psychology and sociology and discussing the current events of the day to our mutual goal of getting a Ph.D.

We also talked a lot about our frustration with the demands of college basketball—the daily practices, some at 5:30 A.M., the frequent long bus rides to games, the missed classes, and the lack of personal

---

LIFE ALONG THE ECONOMIC LADDER

49 percent—*those of us who often find ourselves watching the clock, waiting for the workday to be over*

> *58 percent—those of us earning under $35,000 a year who watch the clock*

> *30 percent—those of us earning more than $75,000 a year who watch the clock*

42 percent—*those of us who say we're only working for the money*

> *50 percent—those of us making less than $35,000 a year who say this*

> *34 percent—those of us making over $75,000 a year who say this*

37 percent—*those of us with concerns about the impact of job stress or working conditions on our health*

> *37 percent—those of us making less than $35,000 a year who say this*

> *43 percent—those of us making more than $75,000 a year who say this*

adapted from Miringoff et al. (2001).

time. From October until March the basketball season dominated our lives, and getting ready for the next season dominated the off-season. The initial thrill of playing college basketball was gradually supplanted by the desire to experience more of life. I wanted to concentrate on my studies, learn transcendental meditation, reconnect with friends, and simply be more spontaneous in enjoying college life. The low point for me occurred at the start of practice for the new season during my sophomore year. There was one moment I remember quite vividly. We were in a practice session performing a conditioning drill in which we would repeatedly run up and down the stairs in our gymnasium. Basketball players from junior high school to the pros perform this drill all the time; it's part of the basketball culture. But for me, on this particular day, it seemed unbearably monotonous. At one point, when I had reached the top of the stairs, I happened to look out the window at the sunlight hitting the red and yellow leaves on a brilliant fall day. The contrast between the awesome beauty outside and the cold, dimly lit gym with its gray concrete stairs sparked an epiphany: My college hoop days were numbered. A few weeks later I called it quits, saying goodbye to college basketball. Although it was clear that I needed to make this decision, it wasn't easy. I experienced great sadness about ending an activity that had been a part of my life since grade school. As a child I had felt destined to play professional basketball—even my initials, NBA, the abbreviation for the National Basketball Association, seemed auspicious to me! Seriously, I knew that I would miss the camaraderie with teammates, the on-campus accolades, the uniforms, and the intense competition. But the prospect of freedom to have a well-rounded life more than overshadowed my nostalgia. Basketball was in my past, and my future was filled with possibilities.

Ron desperately wanted out, too, but felt that he couldn't quit the team because he was attending school on a basketball scholarship. If he quit, he wouldn't be able to stay in school, since his family didn't have the resources to pay his tuition. Basketball was his ticket to an education. Sure, he could've taken out loans or worked his way through, but for him, going into debt was not attractive, and working a job wouldn't have been better than playing basketball, which at least covered all his school expenses. I, on the other hand, was fortunate that my parents could afford the in-state tuition. Although they were far from wealthy, their economic standing gave me the privilege of opting out of an experience that was no longer fulfilling. After I quit, the game remained just that, a "game" that I could play for entertainment or fitness. For Ron it had become a joyless job, simply a means to an end.

The apparent differences in our families' SEP resulted in our varying abilities to make choices. I don't know what happened to Ron. He very well might have achieved the goals we dreamed about, but I'm convinced that in my case my ability to opt out of basketball at will to concentrate on my studies figured largely in my ultimate ability to go to graduate school and develop a satisfying career as a psychologist. It was a pivotal decision point for me.

Without question, choice, flexibility, and control over life's important aspects vary with SEP. The higher your SEP, the more options you have, the more flexibility you have in how you spend your time, and the greater your sense of control over your life in general.[35] This control and empowerment may be especially important for health, since studies suggest that not having them can be deadly. One such study surveyed adult men and women across America to determine the sense of control and power they felt in their lives. They were asked the degree to which statements such as this were true for them: "Many times I feel that I have little influence over the things that happen to me." Scores on these types of questions were used to predict health and mortality over the next five years. *People who felt less power and control over their lives had a 30 percent greater likelihood of dying* during the five years than did those who felt a greater sense of power and control. Powerlessness was also associated with deteriorating health (e.g., illness and disability) among those who survived.[36]

The effects of not having control are striking in the work setting. People in very demanding or physically taxing jobs with little control are at higher risk for heart disease and mortality. Recent research indicates that the truly toxic element of such work strain is not high demand experienced in these jobs; rather, it is the low level of job control that is the deadly ingredient.[37] My former colleague at Harvard, Dr. Jody Heymann, has concentrated on a type of workplace choice that is available to some and not others—time off to care for sick relatives. In her book *The Widening Gap: Why America's Working Families Are in Jeopardy—and What Can Be Done About It*, Heymann documents how many working families are caught between the need to care for children and elderly parents and the demands of work life. Poor and middle-income working families tend to have jobs that do not provide paid sick or vacation leave, jobs in which employees cannot choose flexible times to start or end their workday, and jobs in which employees cannot take off to care for sick family members.[38] Unfortunately, these are the very families that often have more responsibilities to care for elderly parents, parents-in-law, and sick children. The reason for this, as I said before, is that poorer families experience more acute and chronic diseases, with fewer resources to pay for care. Yet, because of a lack of financial resources and job flexibility, working family members often must choose between caring for relatives or earning a living.

In one of the most remarkable and classic studies of the health effects of control, Drs. Ellen Langer and Judith Rodin conducted an intervention in a nursing home to see if boosting perceived control could improve health and well-being.[39] Because the nursing home had multiple floors with little interaction among residents, Langer and Rodin were able to randomly select one floor to receive the intervention and another to serve as the comparison floor. On the surface the intervention seemed very simple. A hospital administrator made a presentation to the residents on the intervention floor that emphasized how much the residents could influence their lives in the facility. They were told, for example, that they could decide how to arrange their personal room, choose where in the facility to visit with friends, and select activities to participate in, such as watching television, reading, or planning social events. They could also decide which of two nights would be set aside for special movie showings. Finally they were offered a small plant as a gift from the facility, which they were free to accept or not. If they accepted the plant, they could decide how to care for it. In reality nothing had changed in the facility— everything the residents were told was already within their power. The administrator simply made their choices and control more salient.

On the comparison floor, residents were given a similar presentation, but the emphasis was on what they were *allowed* to do rather than on what they could *choose* to do. They were *permitted* to visit other residents, their rooms were arranged for their comfort *by the staff*, they were *given* a plant (with no option to reject it) that would be *cared for by the nurses,* and they *were told* which night would be movie night.

After only three weeks the residents given choices reported being significantly happier and more active, and they were rated by the nurses as generally more improved and more active, even though the nurses were unaware of the subtle intervention that had occurred. What happened after eighteen months was startling. Prior to the intervention the average death rate in the entire facility was 25 percent for an eighteen-month period. But in the eighteen months after the intervention, *only 15 percent of the residents on the intervention floor had died, but 30 percent of those on the comparison floor had died.*

As you can see, having a sense of control, being able to choose, and having options and flexibility are certainly critical dimensions of our happiness and health. Since these kinds of options are more available to people at the upper reaches of SEP, do they account for the SEP/health gradient? Science has not as yet provided an answer. But it is an intriguing possibility, since the nursing-home study suggests that even minor shifts in our perceptions of control can be useful.

Off to a Bad Start?

Most of the research on SEP and health focuses on adult economic standing, as scientists determine how an adult's *current* level of education, income, or occupational status relates to illness or death. Yet is it possible that the effects of SEP begin in childhood? Could economic adversity experienced during childhood, or even infancy, set the stage for adult health problems? If so, can removing economic deprivation during adulthood counteract or reverse the negative sequelae of childhood disadvantage?

One of the newest theories suggests that much of the gradient effect in adult health occurs because adults who are less well-off were also less well-off as children. Dr. Clyde Hertzman of the University of British Columbia is one of the scientists saying that early material or psychosocial deprivation contributes to subsequent ill health in adults. There is plenty of evidence to support this notion. First, and not surprisingly, scientists know that children from lower-SEP backgrounds experience more illness and more emotional and behavioral problems

---

LIFE ALONG THE ECONOMIC LADDER

---

37 percent—*those of us worried about teachers' being overly harsh or embarrassing to students*

> *44 percent—those of us with a high-school education or less who are worried about this*

> *23 percent—those of us who are college graduates who are worried about this*

43 percent—*those of us worried that our children are not being sufficiently prepared for the next grade*

> *50 percent—those of us with a high-school education or less who are worried about this*

> *34 percent—those of us who are college graduates who are worried about this*

60 percent—*those of us worried that our children are picking up attitudes and behaviors that go against our values*

> *70 percent—those of us with a high-school education or less who are worried about this*

> *51 percent—those of us who are college graduates who are worried about this*

adapted from Miringoff et al. (2001).

---

than do those of higher SEP.[40] Second, some studies show that children raised in poverty are at greater risk for problems like heart disease as adults, regardless of adult SEP.[41] Finally, the negative effects of childhood economic disadvantage can be cumulative. Adverse health later in life is more likely the longer the child remains disadvantaged.[42] British researcher Dr. David Barker and associates have even speculated that inadequate prenatal care, more prevalent among the poor, may negatively affect the growing fetus in ways that influence the adult's health decades later.[43]

Drs. Carol Ryff and Burt Singer illustrate how cumulative disadvantage over the life cycle may have biological consequences and how social resources can offset the effects of disadvantage. Using data from the Wisconsin Longitudinal Study,[44] they categorized people as either "relatively advantaged" or "disadvantaged" economically at two points in their lives: first when they were eighteen years old, in 1957, and later when they were fifty-two or fifty-three years old, in 1992 or 1993. Relative economic advantage/disadvantage during their younger years was defined by whether their parents were above or below the median income for the state of Wisconsin in 1957. The participants' relative advantage/disadvantage as adults was defined by whether they themselves were above or below the state's group median income.

Investigators sought to determine if a key index of biological functioning—*allostatic load*—was significantly affected by cumulative disadvantage.[45] Allostatic load is a measure of composite biological status that includes such factors as high blood pressure, excess body mass, and high levels of hormones such as cortisol and catecholamines. Higher allostatic-load scores indicate more potentially health-damaging biological impairment. In the Ryff and Singer study allostatic load varied as a result of both childhood and adult economic status. The highest allostatic loads were found in those participants who experienced economic disadvantage during childhood *and* as adults, with the lowest levels seen in those who were relatively advantaged as children and adults. Interestingly, social support was found to counter the effects of childhood disadvantage. Participants who reported strong relationships and close bonds with their parents had lower allostatic-load scores. In fact, close bonds with parents counteracted the effects of lifelong disadvantage. Among people who were disadvantaged as children and adults, those reporting negative social ties with their parents had over three times the allostatic load of those reporting strong relationships with parents.

To this point I have focused on individual risks for illness and death attributable to SEP. But there's another indicator of economic well-being that is just as important and just as deadly. That is the degree of economic inequality that exists between those at the extremes of SEP—the most well-off members of our society and the least well-off. This is the focus of the next chapter.

Chapter 9

Beyond Individual Achievement: Inequality and Race

LIFE ALONG THE ECONOMIC LADDER

74 percent—*those of us who feel that the gap between the rich and the poor is too large*

13 percent—*those of us who feel that the gap between the rich and the poor is just about right*

6 percent—*those of us who feel that the gap between the rich and the poor is too small*

adapted from Miringoff et al. (2001).

In most countries, including the United States, there is a substantial economic gap between the most and the least well-off. This is not surprising since by definition you are looking at extremes of economic standing. But what is less recognized is that the size of the gap varies considerably among countries. For instance, the distance between the most and least well-off is much smaller in Canada, Norway, and Sweden than in the United States, but larger in West Germany. The population of the United States has always been dominated, at least numerically, by middle- and lower-income families. The majority of us fall somewhere within that great middle range of SEP. Yet most of the wealth is concentrated within the relatively small group of individuals and families at the top of the SEP ladder. And that group's share of the wealth is getting larger and larger. For a good portion of the twentieth century, the wealthiest 20 percent of families received about 40 percent of the total income earned by all families. The middle 60 percent received about 54 percent, leaving 6 percent of income for those in the bottom 20 percent. But in the last twenty years or so, things began to change. The wealthiest 20 percent of families now claim about 50 percent of total family income. The 60 percent of families in the middle had a reduction in their share, falling to 49 percent of the total. The remainder, such as it is, goes to the bottom 20 percent. Things got even sweeter for those in the top 1 percent of family incomes during the bull market of the 1990s. This group saw its share of the total jump from 11 percent in 1990 to almost 18 percent in 1999.[1]

Is this really a problem? So what if the rich got richer? Isn't that the way it always is in a free-market economy? Humanitarian and moral reasons notwithstanding, there is a more concrete, self-serving reason for why this expanding gap is indeed a problem: Inequality is bad because it negatively affects the health of everyone. ***When the gap between the rich and the poor is large, everyone's health suffers.*** Here is what science has discovered about inequality.

Inequality predicts life expectancy. In countries where the gap between the rich and the poor is larger, the average life expectancy for everyone is shorter. In West Germany and the United States, where the gap is relatively wide, life expectancy in 1981 was about seventy-three (West Germany) and seventy-four years (United States). But in Sweden and Norway, where the gap is much less, life expectancy was around seventy-seven years.[2]

Inequality predicts mortality within the United States. The health effects of income inequality can be seen in this country by comparing states that have differing income gaps, using something called the Robin Hood Index to measure inequality.[3] As Robin Hood robbed the rich and gave to the poor, the Robin Hood Index calculates just how much income from the prosperous would have to be given to the poor to equalize their incomes. Figure 11 shows the relationship between the Robin Hood Index and mortality in the United States. After taking into account state differences in poverty, mortality is higher in states where the income gap is greater (such as Louisiana, Mississippi,

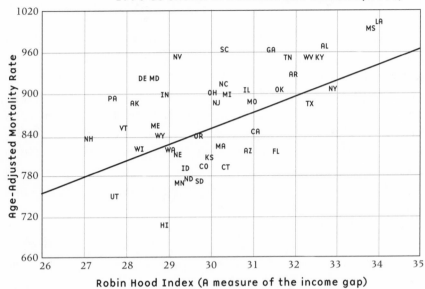

Figure 11: Relationship of Income Inequality
(measured by the Robin Hood Index)
to Age-Adjusted Total Mortality Rates
in the 50 U.S. States (from Kennedy et al.,
1996 as shown in Berkman and Kawachi, 2000)

and New York) than in states where the income gap is smaller (such as New Hampshire, Utah, and Wisconsin). You may recognize that these states differ in more ways than the Robin Hood Index and poverty. In particular, the less equitable states are those with a higher percentage of African Americans who, as a group, have a higher mortality rate than whites. So, it might be argued, the effects could be due to the ethnic differences between the states rather than to the Robin Hood Index. But as the researchers point out, the Robin Hood effect was the same *within* the black and white populations when looked at separately. So the Robin Hood effect is not caused by racial differences in the states' populations.

Changes in inequality predict changes in life expectancy. If income inequality contributes to shorter life expectancies, then you would expect that changes in equality in a region would change the longevity of the residents accordingly. This is exactly what occurs. In countries where the percentage of people in poverty increases over a number of years, the overall life expectancy decreases over the same period.[4] In countries where incomes become more equal over time, there is a corresponding increase in general life expectancy during the same period.[5]

How does inequality kill? Unfortunately, we're pretty far from answering this question with any degree of certainty. But Dr. Ichiro Kawachi of the Harvard School of Public Health, who has conducted much of this research, has some ideas. He says, "Inequality harms health partly because of its deleterious effects on social cohesion. As the rich pull away from the poor, and the social distance grows between the top and bottom of the income distribution, the rich begin to secede from all spheres of civic life. For one thing, they begin to move away to exclusive communities in the suburbs, in some cases to gated communities. They also begin to pull their children out of public schools, and they begin to disappear from civic associations. When the affluent secede from civic life in these ways, two things happen. First, the tax base of communities dwindles when the more affluent move away, which leads to a degradation of public goods, such as the quality of public schools, and the quality of civic amenities and services. Secondly, the affluent who remain have a disproportionately strong voice in local politics, and are often strong advocates of cutting taxes that finance public goods, further exacerbating the problem. In other words, inequality tears at the social fabric." As support for his ideas, Kawachi has demonstrated a strong correlation between growing inequality and the decline in social capital—that is, states that tolerate vast economic inequalities are those that also have high degrees of civic mistrust.[6]

The Color of Inequality

The role of race and ethnicity is an issue that frequently arises when discussing SEP and inequality. Historically, using race and ethnicity—defined as blacks, whites, Latinos, American Indians, Asian Americans, and Pacific Islanders—is one of the principal ways we categorize and group each other in the United States and to make social judgments.[7]

Nearly forty years since the landmark civil rights legislation of the 1960s, race and ethnicity are still used quite frequently by many people to determine everything from where they live, where they worship, whom they marry, to which funeral home they use for their loved ones.[8]

Not only do race and ethnicity shape many of our life experiences, they can be powerful predictors of longevity. Although many differences in health and longevity are evident between racial and ethnic groups, perhaps the most striking example is the health difference between blacks and whites. Compared to whites, blacks suffer higher death rates from nearly every illness, including heart disease, cancer, diabetes, cirrhosis of the liver, and HIV/AIDS, as well as from homicide.[9]

A key question, then, is why? Why would one's race have such profound effects on longevity? This question is the source of intense scientific investigation across the country and has become a priority of government health agencies such as the National Institutes of Health and the Centers for Disease Control and Prevention. So what do we know? We know from the government's massive Human Genome Project and other genetics research that genes are an unlikely source of the differences. When it comes to genetics, we're all pretty much the same. According to the National Genome Research Institute, all humans share about 99.9 percent of their genes, and geneticists have long believed that genetic differences between *individuals* is much more significant than difference between groups.[10]

The difference between blacks and whites in the performance of behaviors that relate to health—such as smoking, physical activity, and diet—are thought to play a role in health differences. SEP is certainly a key culprit, given the large economic gap between the groups. Although this economic gap has been closing in recent years, blacks still have, on average, lower incomes and educational levels than do whites.[11]

Given SEP differences between blacks and whites, you might suspect that SEP accounts for the gap in health between those groups. And indeed it does. For some illnesses, if you statistically account for group differences in SEP, the race/health gap goes way down, and in some cases it is completely eliminated. A study by the Centers for Disease Control and Prevention is a case in point. In a large study of black and white adults, about 38 percent of the gap in deaths between the groups was due to group differences in SEP. Another 31 percent of the gap was due to well-known risk factors such as physical inactivity, smoking, high blood pressure, high cholesterol, and excessive body weight. The remaining 31 percent of the difference could not be accounted for.[12]

So SEP is an important part of the black/white health gap, but what about that 31 percent that could not be explained? Could that have something to do with SEP as well? I think it does, but it may not have to do with *individual* SEP. There is some indication that even when individual SEPs are the same, blacks fare worse than whites in terms of health, especially those blacks at the lower end of the economic spectrum. That is, poorer blacks tend to have higher death rates than do poorer whites.[13]

What could be happening? SEP could still be playing a role, but perhaps not individual SEP so much as the collective SEP of the neighborhoods where poorer blacks reside. Poorer blacks and poorer whites generally live in fundamentally different types of neighborhoods. Poor blacks tend to live in neighborhoods where most of their neighbors are also poor, whereas poor whites live in areas where the poor are not in the majority.[14] In other words, poorer whites are more economically integrated with others.

Why would this matter for explaining health differences between poor blacks and whites? It's important because neighborhoods where more poor individuals live are, not surprisingly, also those neighborhoods with the highest death rates. But what *is* surprising is that these neighborhoods are unhealthy ***even for the nonpoor who live there.***[15] Deaths from all causes are higher in these neighborhoods, not just causes typically associated with poor areas, such as homicide. And again it is not just the poor who are at risk—a middle-class person living in a poor neighborhood is at greater risk for death than a middle-class person who lives in a nonpoor area is. We don't as yet know why this is the case, but whatever the reason, blacks are more exposed to it than are whites.

In sum, it is clear that SEP is a big part of the race/health puzzle, and it can exert its influence in some surprising ways. But beyond the race question, one important point is that SEP is not just about individuals and their learning, earning, and achieving. It is also about the collective SEP of the neighborhoods and communities in which we live. When the overall economic well-being of communities is low, we all can suffer, regardless of our personal SEP.

CONCLUSION TO PART IV

When I was at NIH, I served on a committee that was developing a plan to increase funding for research on health disparities, which would provide information to better understand and eliminate racial and economic differences in health. At one of the meetings, someone remarked that perhaps SEP did not warrant such concern and that NIH should focus on things that science can study and actually do something about. This person was unaware that NIH actually funds quite a bit of research on SEP and health. And I hope from this chapter it is evident that the health effects of SEP can be studied using standard scientific methods. What is less clear is what should be done in response to the findings. Given what we know about SEP, what steps should we take to enhance longevity?

The nature of economic influences on health raises questions on both the individual and societal level. On an individual level, should we strive to get more education, earn more money, or apply for jobs that allow more control and decision-making ability for the health benefits that might accrue? That is, should we work to lessen the impact of the gradient by improving our position on it? In many respects the answers to these questions for many of us would be yes, even if the SEP/health gradient did not exist. Moving up a rung or two on the SEP ladder to make our families and ourselves more financially secure is very much a given goal in our society. There is an intrinsic value to obtaining more education or a job that offers more flexibility, control, and remuneration. The fact that achieving greater economic security or education carries a health benefit with it is icing on the cake, not necessarily the motivating influence.

A more complex challenge raised by the gradient is that it compels us to look beyond ourselves and our individual needs—away from "*my* health and well-being"—to the health needs of society as a whole, with an emphasis on "*our* health and well-being." The most pressing issues raised when I consider "our health" from the perspective of the SEP/health gradient are the health effects of economic inequality and the disproportionate health burden experienced by those at the lower end of SEP. Our collective health is very much tied to the economic distance between the wealthiest and the poorest in our society. The tremendous disease burden borne by the least well-off among us is inconsistent with the principles of equality and fairness to which we as a society aspire. Therefore a goal of lessening the impact of economic inequality and low SEP is one that combines both enlightened self-interest and altruism. But there is only so much we can do as individuals to achieve this goal. Health-policy changes are required. Does this mean that leaders in local, state, and national governments should be interested in the SEP/health gradient and economic inequalities? Absolutely. Think about it. If biological researchers were to discover a virus that is associated with nearly every cause of death and disability, resulting in untold suffering and medical costs, altering that virus would be a national health priority. Economic inequality and low SEP are the social equivalents of that hypothetical virus, and altering *their* impact should be a national health priority. We can decide to improve the lives of those most negatively affected by the gradient—those working families and individuals of lower SEP. Although SEP also affects the middle and upper classes, persons of low SEP bear its full weight—poor-quality or no health care, unhealthy neighborhoods, factors that promote health-damaging behaviors, and chronic stress, to give just a few examples. We can also take action to reduce the economic gap between those at the very top and those at the bottom of SEP. When this gap is small, we all benefit. Surprisingly, the public-policy changes required to achieve these ends would not mean radical departures from the norm, only a change in emphasis and priorities. A recent report entitled "Improving Health: It Doesn't Take a Revolution,"[16] outlines modest steps we can take as a society to alter the effects of the SEP gradient. Here are a few:

- Invest in young children through policies that explicitly recognize how important early development is throughout life, such as improved parent training and preschool-education and -nutrition programs.

- Provide services and opportunities that seek to close the gap between the prosperous and less prosperous, including improved low-income housing; community education, nutrition, job training, and disease-prevention programs; and better access to quality health care for all.

- Improve work environments by involving employees in decision making, offering flexible leave policies and work schedules, and offering greater opportunities for career development.

- Provide stronger support for community efforts aimed at facilitating economic development and empowerment, increasing civic participation, and building stronger social networks to counter the effects of lower SEP.

- Reduce economic inequities through tax, transfer, and employment policies. Examples include increasing the Earned Income Tax Credit (which reduces the tax burden on workingpeople with very low incomes), the minimum wage, and unemployment compensation.

- Assess the health impact of national financial and social policies. For example, what are the health implications of substantially decreasing the tax burden on the wealthiest 5 percent of Americans (as we have done in recent years)? Does this increase or decrease inequalities? What is the impact on our nation's health as a whole?

We don't have to start a social revolution or abandon our free-market economy to do the right things to reduce health disparities. Incremental changes over time can have big health payoffs.

PART V

FAITH AND MEANING: EXISTENTIAL, RELIGIOUS, AND SPIRITUAL DIMENSIONS OF HEALTH

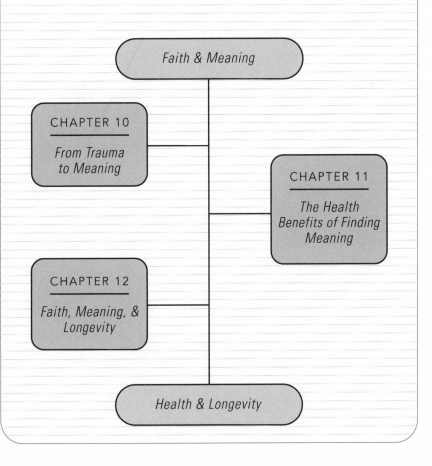

Faith & Meaning

CHAPTER 10

From Trauma to Meaning

CHAPTER 11

The Health Benefits of Finding Meaning

CHAPTER 12

Faith, Meaning, & Longevity

Health & Longevity

*. . . there is nothing in the world, I venture to say, that would
so effectively help one to survive even the worst conditions as
the knowledge that there is meaning in one's life.*
—Viktor Frankl, *Man's Search for Meaning*[1]

The Holocaust stands out in world history as among the most heinous treatments of one people by another. Survivors, "refugees from oppression," as William B. Helmreich refers to them in *Against All Odds*, endured sustained and systematic torture.[2] Millions died. Helmreich spent six years exploring the lives of the 140,000 Jewish refugees who had spread out across the United States by 1953, and his highly acclaimed work is comprehensive and revealing. Helmreich conducted 170 personal interviews to find out how people who had survived the "cataclysmic events" of the Holocaust pieced together the shreds of their lives to carry on. Accounts of how survivors fared after the war are as individual as their lives were before the Holocaust. Many achieved success financially and personally, becoming involved in family and community life. Helmreich discovered ten attributes common among the survivors who went on "to lead positive and useful lives following the war," and he gives his readers great insight into how they prevailed. Interestingly, two of the ten characteristics that he observed, I have discussed in previous chapters: optimism and social support. A third is the focus of this part of the book—finding meaning. Helmreich explained that after accepting that they had somehow survived when others had not, an important aspect of achieving a sense of well-being was the survivors' ability to translate their "good fortune into a concrete reason to carry on." In essence the survivors who adapted well were able to interpret their survival in a way that gave purpose to the rest of their lives.

The challenges we endure obviously cannot compare to the long-term brutality of the Holocaust, but ours can nevertheless seem overwhelmingly painful to us. And we must somehow marshal our resources to survive. But how? How do we come to terms with experiences so devastating that they threaten to destroy our understanding of the world? Events that irrevocably change our lives, placing us on a path that we neither anticipated nor desired? Events that shatter long-held assumptions? Events of such magnitude that they threaten even the most optimistic explanatory style—such as receiving a terminal medical diagnosis, experiencing the death of a child, being held in captivity as during the Holocaust and slavery, or losing a loved one during the frightful terrorist attacks on September 11, 2001. These kinds of events transport us beyond simply wanting to identify a cause for what has happened, beyond desiring to attribute responsibility to someone or something. These are circumstances that move us to ask the more existential *why* question, "Why did this happen?" or, more pointedly, "Why did it happen *to me or to my loved one?*" Every day someone raises these kinds of questions in different forms: "Why did my spouse have to die?" "Why did my child have to die?" "Why did my marriage have to end?" "Why did I develop cancer?"

In the chapters that follow, I explore research on, and personal stories about, the *why* question, which in its essence is a search for meaning. Finding meaning is the process of gaining a deeper comprehension or understanding of events that are unquestionably aversive. But finding meaning is more than gaining understanding. It also involves discovering benefits, experiencing growth, or even thriving after major life reversals. I chose the area of finding meaning as an example of the existential/religious/spiritual aspect of health because, as you will see, it effectively encompasses many relevant themes.

Reynolds Price is one of those persons who experienced disaster but who was ultimately able to derive meaning from it. The award-winning author and distinguished Duke University professor was at the height of his writing career when, in 1984, he was diagnosed with a malignant spinal tumor. He survived, and, although confined to a wheelchair, in his book *A Whole New Life* he wrote,

> So *disaster* then, yes, for me for a while—great chunks of four years. *Catastrophe* surely, a literally upended life with all parts strewn and some of the most urgent parts lost for good, within and without. But

if I were called on to value honestly my present life beside my past—
the years from 1933 till '84 against the years after—I'd have to say
that, despite an enjoyable fifty-year start, these recent years since full
catastrophe have gone still better. They've brought more in and sent
more out—more love and care, more knowledge and patience, more
work in less time.

Price emphasizes the words "disaster" and "catastrophe," no doubt
to illustrate, in unmistakable terms, the magnitude of the impact that
the illness had on his life. Yet out of the wreckage of cancer and disability a new life emerges, one that he says has "gone still better" than
the fifty years before. He alludes here, and elsewhere in his book, to
improvements in both his relationships and his work productivity.
Price's story illuminates in a single experience what scientists are documenting on a larger scale: People frequently experience a dramatic
and unexpected upside from adversity and trauma.

Research on finding meaning after a crisis is not new, but it is just
beginning to blossom. To me it is one of the most intriguing aspects of
the new health science, because it relates to an age-old human concern—the meaning of suffering. Religious and spiritual leaders have
dealt with this for centuries. Indeed, religion and spirituality are key
routes to finding meaning for many people, and scientists are now
systematically exploring just how effective religious practices actually
are during trials and tribulations.

It turns out, as research is demonstrating, that emotional turmoil
is only part of the adversity experience. For many of us adversity can
have a paradoxical, unforeseen, yet undeniable value—contributing
to both emotional well-being and, often, extended life.

Chapter 10

From Trauma to Meaning

Trauma in the Greater Scheme of Things

In October 2000 Carol Muske-Dukes was ready to publish her latest novel, *Life After Death,* which the *Los Angeles Times* would later describe as a tale "about the fate of a young couple when the husband dies unexpectedly in a tennis game. The bulk of the novel revolves around the dialogue that continues between two lovers even after one of them dies."[1] On October 9 the life-imitating-art cliché became all too real for Muske-Dukes. Her husband of seventeen years, a fifty-five-year-old actor, died of a heart attack on a business trip—while playing tennis. He had no history of heart disease. "There was no pain, no illness, no hint that anything was wrong," Muske-Dukes recalled. With his sudden death her world turned upside down. She goes on to say, "Suddenly, the person who has defined your galaxy is gone. The solar system shifts. You are no longer in the same place vis-à-vis the sun. Your compass no longer points north. It is spinning."[2]

For Muske-Dukes, and for thousands of other people, the tragedy of a devastating loss is truly a transformative event. Many survivors of tragedy divide their lives into two parts—before and after. Traumas challenge our long-held, often lifelong assumptions and core beliefs about the world. In the chapter on expectations and explanations, I discussed how our core beliefs or worldviews can determine whether we interpret events in optimistic or pessimistic ways. But our assumptions and theories about the world have an existential quality that extends beyond optimistic or pessimistic styles. As psychologist Dr. C. Murray Parkes has noted, our "assumptive world is composed of a strongly held set of assumptions about the world and the self which is confidently maintained and used as a means of recognizing, planning and acting."[3] Research demonstrates that tragedy—an unwanted, unexpected turn of events—shatters our assumptive world, leaving us with a sort of existential void. Life makes less sense than it did before the tragedy. Our success or failure in reconstructing a new worldview—one that makes the untoward event understandable—plays a significant role in determining our future well-being.

With extremely traumatic experiences, what happens to most of us is analogous to the process of learning about the world as infants and young children. Psychologists know that the very young learn about the world through two processes: *assimilation* and *accommodation*. With assimilation children attempt to make new experiences fit into categories or cognitive "schemas" that already exist. Imagine that cognitive schemas are little file folders that we have in our heads to categorize what we know. When faced with a new stimulus, we will try to file it away—that is, assimilate it—within an existing schema. To a child with limited schemas, an airplane may be a "bird," a strange man may be "daddy," and a bus may be a "car." When new experiences do not easily meld with existing schemas and when new learning occurs, accommodation begins. With accommodation the child learns to develop additional schemas, such as a new schema for "man" that is different from the schema for "daddy." For the very young, accommodation is a transformative experience, permanently changing and expanding their worldview, providing new information with which to interpret experiences.

Like children, adults also have cognitive schemas. Many psychologists believe that adult schemas include our beliefs, assumptions, expectations, and theories about the world.[4] In fact, adult schemas include everything that we know, believe, assume, or think is relevant and important about a particular topic. Schemas help us sift through the barrage of information to which we're constantly exposed. Like young children, adults also must determine if new information can be assimilated into existing schemas or if it must be accommodated by changing a schema or creating a new one.

We unconsciously ask, "Is this new information congruent with what I know about this category, or is it incongruent?" "Does this new information require me to change?" The thing to remember about schemas is that they *do not* change very easily. Assimilation is the rule, and accommodating by developing new schemas is avoided. We will quickly reject and forget information contradictory to our schemas but accept at lightning speed and remember verbatim information that supports our ingrained beliefs. We want to keep believing what we already believe.[5] For example, let's say you have a schema about airplanes that includes the theory that planes are unsafe. You're likely to pay little attention to the fact that thousands of planes take off and land successfully each day around the world. You ignore that, because it is inconsistent with your schema about planes. But if there's an accident, your schema has the data it needs to justify its existence. "Aha!" you might think. "I knew planes weren't safe!"

Why do we hold fast to our assumptions and beliefs even in the face of contradictory information? According to psychologist Dr. Ronnie Janoff-Bulman of the University of Massachusetts, the author of *Shattered Assumptions: Towards a New Psychology of Trauma,*[6] "Our conservatism in maintaining schemas derives from our fundamental need for stability and coherence in our conceptual systems. Generally, our schemas serve us well. They are constructed and solidified over years of experience and, as stable knowledge structures, provide us with the necessary equilibrium to function in a complex, changing world."[7] Janoff-Bulman believes that the reason traumatic events are so difficult for us is that they challenge and threaten our schemas about the world. We assume that the world is generally a good place, and even though bad things happen, we assume that they won't happen to us.[8] We believe that the world is just and meaningful, that good and bad events are somehow distributed with fairness. And even though we have plenty of evidence to the contrary (e.g., random acts of violence, infant deaths and disabilities, families with more than their fair share of hardships), we want to believe that people generally get what they deserve, what psychologists call the "Just World Theory."[9]

But then the unthinkable happens. We develop cancer, have a serious accident, lose a child or spouse, or are the victim of crime. We are confronted with incontrovertible evidence that our basic assumptions about ourselves and the world are incorrect. When her husband died unexpectedly, Muske-Dukes had to assimilate the experience into her established schema about herself, her partner, and their planned lives together. This process of schema change is extraordinarily difficult. A surviving spouse cannot easily assimilate the new, tragic experience, yet it is too salient, too unrelenting, to dismiss. I asked Janoff-Bulman to explain why it is so hard to assimilate events that violate our assumptions. She said, "Traumatic events are too emotionally powerful to discount or ignore. They dramatically challenge the very assumptions that had enabled victims to approach their daily lives with a sense of safety and security. We are generally resistant to changing our schemas, particularly our most fundamental theories about the world and ourselves, for these provide the foundation upon which our other beliefs are built. Change at this level threatens the breakdown of the entire system, as major cracks in the foundation of a house endanger the stability of the entire structure. Yet victims are suddenly all too aware that their old, comforting theories can't account for their traumatic experience. Their coping task is monumental, as they struggle to maintain the coherence of their psychological world while working to integrate their potent, adverse experience. Ultimately, trauma victims strive to rebuild and reshape their inner world so that their new fundamental assumptions can acknowledge the reality of their traumatic events yet provide a view of the world that is not wholly threatening. Psychologically, this is an incredibly tall order."

For some victims the stress created by the incongruity between the trauma and their assumptions leads them to completely change their long-held assumptions about the benevolence of the world and even their own self-worth.[10] To make the traumatic experience fit, they now assume that the world is not a just and benevolent place or that they are somehow not deserving of a joyful life. Such a reassessment can lead to a downward spiral of pessimism and cynicism, and potentially to ill health. But research is discovering that for many others the reconciliation of the conflict between assumptions and reality does not have to follow this more negative course. Many survivors have found another way out of the cognitive dilemma—they find meaning and grow from adversity.

What Does It All Mean?

In discussing her husband's sudden death, Muske-Dukes expressed both regret and a new perspective on life: "We were moving so quickly that it seems to me we never got a chance to really celebrate the extraordinary richness of our lives together. I mean, we did it, but in such a casual way. It's a great lesson to me now. . . . I realize that I must stop to appreciate what is immediately in front of me at a given moment."[11]

This is a sentiment shared by Linda Ellerbee. You know from previous chapters that this outstanding newswoman is a longtime breast-cancer survivor. She also talked to me about the new meaning that took shape in her life because of her diagnosis.

A good friend of Ellerbee's who was HIV positive told her that good things would come to her from having a life-threatening disease. She said that she thought something like "Yeah, right," but later she understood how right he had been. "Good things did come to me," she said. "A better sense of proportion about my life, all the corny stuff—you do take more time to smell the flowers, but you have to keep reminding yourself." Like many survivors, Ellerbee finds meaning in simply being alive. She said, "Since I've had cancer, every summer on my birthday, in August, I load up my backpack and I go out for a week alone in the wilderness. That week is a time for me to shut up and listen. To see the beauty in a blossom or hear the beauty in the wind and the trees or feel the beauty from sleeping under the stars at night, and to be grateful for the year I have just had. To be grateful that I am awake, alive, and healthy on the top of a mountain, having survived cancer for my birthday."

Obviously neither Muske-Dukes nor Ellerbee wished for adversity. None of us does. It comes as an unwanted, unwelcome guest, invading our life space and draining our emotional and physical resources. How can we get back on track and prevent life reversals from claiming complete victory? A key to recovery is our ability to find meaning in the event. This entails uncovering an acceptable answer to the *why* question. As I explained, finding meaning may involve assimilating the event into our existing schemas, creating new worldviews to accommodate it, or even redefining who we are.[12]

Reynolds Price, for example, described in stark terms how essential it was for him to come to terms with his new circumstances and how beneficial it would have been if he had more quickly redefined himself after his crippling cancer:

> The kindest thing anyone could have done for me, once I'd finished five weeks' radiation, would have been to look me square in the eye and say this clearly, "Reynolds Price is dead. Who will you be now? Who *can* you be and how can you get there, double-time?" Cruel and unusable as it might have sounded in the wake of trauma, I think its truth would have snagged deep in me and won my attention eventually, far sooner than I managed to find out myself.[13]

"Reynolds Price is dead. Who will you be now?" These words that Price says would have helped him during his lowest point are jolting, yet they effectively communicate a major step toward finding meaning that may necessitate letting go of an old self or old assumptions and re-creating and redefining who you are now. Some people actually come through a traumatic experience with a perception that their lives benefited from it, that aspects of their lives are not only different but are somehow improved as a result of adversity. It has been estimated, for example, that as many as 50 percent of people who experience life crises report some benefits from them.[14]

Representative Tom Osborne is one of those people. A fourth-generation Nebraskan, he was sworn in as a member of the U.S. House of Representatives in January 2001. Following college graduation he played three seasons in the National Football League, before an injury ended his career. He told me, "At the time it was difficult to see, but as I look back on it, it was probably the best thing that ever happened." Without the injury Osborne said that he might have had another ten years with the NFL. But professional football wasn't as lucrative a profession when he played. He added that he probably would have retired in his early thirties, without a plan. Because of the injury he returned to the University of Nebraska, where he met his wife and earned a master's degree and a doctorate in educational psychology. He worked as a graduate assistant to the football coaching staff, which led to a coaching career of his own. He coached the Cornhuskers from 1972 until 1997, leading them to three national championships. Osborne probably could not have imagined such a wonderful outcome from that injury. He sums up a common feeling: "Adversity is not always your enemy. Adversity in many ways is your friend, and most growth and progress occurs through adverse circumstances. It depends on your orientation."[15]

Matt Varney, the young Columbine High School student whom I mentioned in the chapter on disclosure, learned the hard way to include finding meaning in his approach to healing. Part of his "healing backwards" was trying to help others find meaning before he explored his own feelings to understand what the tragedies meant to him. He said, "I stayed extremely busy and therefore didn't have time to realize what my own feelings were." The suicide of his friend Greg forced Varney to reassess what was going on around him in order to keep his footing. "It wasn't healthy for me just to stay busy all the time. When Greg killed himself, all of that blew up straight in my face, and I had to start all over," Varney said. Now he knows that "I want to make [the Columbine students'] loss meaningful to my life and I do that by speaking out and carrying on their legacy." Remembering his basketball buddy, Varney says that he wants to use his gift for speaking out, which manifested after the Columbine tragedy, to expose other young people to the threats of suicide. He has finally found meaning in what happened and found an anchor for his own future as a result. Varney says that he's proud to be a resource of hope to other people.

Dr. Crystal Park of the University of Connecticut, an expert in research on finding meaning, describes it as a process, much like the experiences of Varney, Price, and others would suggest. She told me, "Finding meaning involves the active, but not necessarily deliberate, attempt to reevaluate or see the stressful experience from a different angle, so that it becomes less discrepant with your views about how the world works or your desires about what you *wanted* to have happen. This process of reappraisal can make the event less threatening to your cherished beliefs and goals." Park warns, however, that sometimes such a reappraisal is not always effective, especially for events that are so dramatic that people have to alter their long-held beliefs and goals. But this is not always bad. "Using the experience as a catalyst for positive changes in one's life is one of the more adaptive strategies that a person can take to make meaning," she said.

Dr. Shelley Taylor of the University of California at Los Angeles has written extensively about the experience of finding meaning in traumatic events, which she believes is a key component of what she calls "cognitive processing" of the event. Cognitive processing involves several phases, including actively thinking and ruminating about the traumatic event, paying attention to the thoughts and feelings that arise, and assessing the implications the event has for life and the future.[16] Although in the chapter on disclosure I talked about rumination in a negative light, it is actually the first phase of coming to terms with adversity. It is only a problem if it continues unabated and no action is taken to move beyond it. New and more uplifting themes may emerge as we move beyond rumination to assess the implication of the event for our future.

Taylor found that a surprisingly high number of the 78 women in her breast-cancer study reported positive outcomes associated with their illnesses. Many made comments like these:[17]

"I feel as if I were, for the first time, really conscious."

"I have much more enjoyment of each day, each moment. I am not so worried about what is or isn't or what I wish I had. All those things you get entangled with don't seem to be part of my life right now."

"You take a long look at your life and realize that many things that you thought were important before are totally insignificant. That's probably been the major change in my life. What you do is put things into perspective. You find out that things like relationships are really the most important things you have—the people you know and your family—everything else is just way down the line. It's very strange that it takes something so serious to make you realize that."

Taylor's patients are not unique in their experience. Research has now documented this remarkable ability to find meaning and benefits from adversity among many people who experience life-threatening illnesses, bereavement, divorce, natural disasters, and even exposure to war.

Cancer. Scientists have documented that many cancer patients report a new, more positive perspective on life as a consequence of illness. In a summary of a number of studies on this topic, the following conclusion was reached: "Although associated with an intense crisis, cancer concurrently generates significant gains in life appreciation arising from the confrontation with mortality, sickness, and the struggle for health. Not only do survivors appear to become more satisfied about their lives as a whole, but they become more accepting of themselves and often find renewed interest in religion while they reflect on an improved quality of life for the present."[18] In the chapter on disclosure, I talked about Margie Levine, the feisty woman with that rare form of lung cancer. She explains that spirituality played a big role in her healing. "I always tell people that if you get diagnosed and you don't have a spiritual connection, you will soon, because it really opens you. It opens you fully."

Stroke. In a study of patients who had suffered a stroke and their family caregivers, *50 percent of the patients and 45 percent of the caregivers were able to find some meaning* in the experience. Many of the patients reported that after the stroke they were able to appreciate life more, learned to slow down, became more compassionate, and appreciated family and friends more. The most frequent response among the caregivers was that they appreciated the patient more than before he or she became sick.[19]

Bone-Marrow Transplantation. Bone-marrow transplantation is potentially lifesaving for many patients with malignancies, but it is also a very aggressive treatment that can be emotionally draining, physically taxing, and associated with a high risk for death.[20] Yet after this procedure many patients also report positive effects, which occur alongside the negative symptoms. One study found that among 90 bone-marrow transplant patients,

- 59 percent reported developing a new philosophy of life
- 47 percent reported a greater appreciation of life
- 71 percent reported making changes in personal characteristics and attributes, such as becoming less selfish or more outgoing

- 52 percent reported improved relationships within the family
- 39 percent reported finding help and support from friends

HIV/AIDS. Recent research with persons who are HIV positive or have AIDS demonstrates that personal growth can accompany the adverse concomitants of the infection.[21] In a New York study of 54 ethnically diverse women with HIV/AIDS, *83 percent reported at least one positive change in their lives that they attributed to the illness.*[22] These positive changes included those listed below:

- adopting a healthier lifestyle—especially giving up drug or alcohol abuse, smoking, and sexually risky behavior—and eating a healthier diet
- gaining strength through spiritual and religious faith
- improved interpersonal relationships with their children, other family, and friends
- improved view of themselves as stronger, more caring of others, and more responsible
- greater appreciation for the little things in their lives and greater value for life in general
- more goal-directed actions, including the pursuit of education, becoming involved in AIDS advocacy or providing care for others, seeking a different type of work, or quitting work to pursue long-standing interests

Preterm Birth. In the book *Infants in Crisis: How Parents Cope with Newborn Intensive Care and Its Aftermath,*[23] Drs. Glenn Affleck, Howard Tennen, and colleagues describe in impressive detail the psychological and social challenges parents face when an infant is born prematurely. Many of these parents express feelings of victimization and despair, but this is only part of their experience. Approximately three-quarters of the mothers admitted asking "Why me?" Of those, 42 percent stated that they had developed an answer. Their answers varied, but the most frequent response related to God's will (e.g., "God selected me to care for this special baby" or "This is probably one of the most important things God will ask me to do"). Even more impressive was that *nearly 75 percent of the mothers cited at least one benefit from the preterm birth,* including increased empathy, better relationships with family and friends, and the feeling that the child was now more precious to them.

Illnesses are not the only adverse life events out of which people derive meaning or find benefits. Following natural disasters such as Hurricanes Andrew and Hugo, children were found to grow closer to their parents, siblings, family members, and neighbors and to show increased concern and caring for others.[24] Many women who experience marital separation show more positive self-esteem, greater feelings of freedom and control, and larger social-support networks than do married women.[25]

The personal stories and research findings described in this chapter indicate that while tragedy can be debilitating, it can also create an opportunity for growth and the discovery of meaning. Although this in itself is an important end, it does not necessarily indicate that those who find meaning are, or have the potential to become, healthier than those who do not find meaning. Does finding meaning produce emotional and physical health benefits? The next chapter will explore that question.

Chapter 11

The Health Benefits of Finding Meaning

R esearch indicates that finding meaning after experiencing adversity can be valuable, especially if it helps us assimilate the stress and trauma and move past it. Yet the benefits of finding meaning go beyond a sense of coherence. Research now suggests that people who find meaning are able to adjust better to adversity over the long term, as indicated by their lower levels of depression, anxiety, and stress. They also experience better immune status, lower risk of certain disease conditions, and increased longevity.

Finding meaning leads to better adjustment after combat exposure. Dr. Carolyn Aldwin and colleagues at the University of California at Davis studied the effects of combat on veterans of World War II and the Korean War.[1] She was interested in identifying factors that determined whether combat exposure early in adulthood would result in symptoms of posttraumatic stress disorder later in life. Aldwin discovered that most of the men viewed military experience and combat exposure as having both negative and positive effects. More than 90 percent said that military service helped them learn teamwork and cooperation, that it broadened their perspectives, and that it taught them how to cope with adversity. Most also acknowledged the undesirable effects, such as being separated from loved ones and the disruption of their lives or careers. Interestingly, however, the men who were able to find the most positive effects from serving in the military and experiencing combat were the least likely to experience combat-related stress symptoms. The more the men viewed their military experience as undesirable, the more they experienced stress symptoms. These findings remained significant after accounting statistically for the degree of combat exposure.

Finding meaning leads to better adjustment to bereavement. Researchers at the University of Michigan conducted an extraordinarily rigorous prospective study of people coping with the loss of a family member.[2] The participants were relatives of patients who were terminally ill and receiving hospice services. The family members were interviewed on average about three months before their relative died and then again periodically for up to eighteen months after their loss. Assessments of various indicators, including depression, anxiety, and stress, were conducted to determine how the participants were adjusting. An interesting feature of this study was the explicit examination of two dimensions of meaning: making sense of the loss and finding something positive in the experience. To assess whether they were able to make sense of the loss, participants were asked, "Do you feel that you have been able to make sense of the death?" To ascertain positive implications, interviewers explained that sometimes people who lose a loved one find some positive aspect in the experience. Interviewers gave examples, such as that some people feel they learn something about themselves or others. They asked, "Have you found anything positive in this experience?"

To the "making sense" question, two of the responses were these:

"My basic attitude to life was that there's a beginning and an end, and it's going to happen to one or the other of us sooner or later, and you have to cope with it. That's all. There's nothing you can do to prevent these things from happening. They're a part of life."

"I think my father's illness was meant to be, and that was God's plan. . . ."

In response to the "benefits" question, responses included these:

"Yes, I found a growth and a freedom to give fuller expression to my feelings, or to assert myself, to do things that I want to do."

"I have learned and seen a lot of positive things in people—they just glowed. It was nice to get that blessing in disguise."

"It was an end to her suffering."

People who were able to make sense of the loss and those who acknowledged benefits showed better adjustment in the months after the death than did those who were less able to find any meaning in their loss. But the two forms of meaning were not equal in their effects. The distress-reducing effects of making sense seemed to diminish about a year after the loss. However, being able to find some benefit following the loss had more enduring effects and was strongly associated with better adjustment eighteen months after the loss. The findings remained significant even after accounting statistically for optimism/pessimism, level of distress before the loss, age of the deceased, and religious belief systems.

Finding meaning leads to lower stress hormones in breast-cancer patients. At the University of Miami, researchers conducted a randomized-treatment study involving stress-management training for women with early-stage breast cancer, who were assigned to either an immediate treatment group or a waiting list for future treatment that served as a control group.[3] The treatment involved ten weekly sessions of group-based cognitive-behavioral stress management, which consisted of teaching the patients relaxation techniques, assertiveness training, social-support utilization skills, anger management, how to control negative thoughts, and other stress-coping skills. Women in the treatment group exhibited significantly lower levels of the stress-related hormone serum cortisol compared to women on the waiting list. Treatment participants also reported an increase in the number of benefits they perceived to be associated with having breast cancer. No such change was found in the control group. Remarkably, further statistical analyses revealed that the reduction in cortisol levels in the treatment group was *a direct result* of the increases in finding benefits from the cancer. That is, the treatment led to reductions in cortisol *because* the treatment helped the participants find benefits.

Finding meaning leads to less risk for recurring heart attack. To explore the effects of perceiving benefits during recovery from a heart attack, researchers at the University of Connecticut School of Medicine interviewed 287 male heart-attack survivors approximately seven weeks after the attack.[4] A portion of these men were contacted again eight years later to determine who had experienced or died from a second heart attack. The follow-up interviews consisted of asking the survivors, among other things, if they perceived any benefits or gains from the experience. During both the seven-week and eight-year interviews, nearly 60 percent of the survivors reported some gains or benefits from the heart attack. The three most frequently cited benefits were as follows:

- a renewed commitment to a healthier lifestyle (e.g., exercise, smoking cessation)
- an increase in activities to enhance enjoyment of life, such as living a less hectic pace and taking more vacations
- a change in philosophy of life, such as valuing home life more, becoming content with one's lot in life, living each day as it comes, and renewed faith in God

The results showed that patients who were able to find benefits from the first heart attack were significantly less likely to have had a subsequent heart attack or to experience other illnesses or symptoms in the ensuing eight years.

Finding meaning leads to better immune functioning and lower mortality among the bereaved. As part of the Multicenter AIDS Cohort Study based at UCLA, scientists interviewed 40 HIV-positive men who had recently lost a close friend or partner to AIDS.[5] They wanted to determine if cognitive processing of the death and the discovery of meaning related to it were predictive of immune functioning and mortality over the course of three years. Cognitive processing was defined as:

- actively or deliberatively thinking about the death
- thinking about the person who died or one's relationship with that person

- thinking about one's own life, mortality, or illness
- thinking about one's actions toward the person before the death

The discovery of meaning was defined as a major shift in values, priorities, or perspectives in response to the loss. Such shifts included:

- a greater appreciation for loved ones
- an enhanced sense of living in the present
- a perception of life as fragile and precious
- a commitment to enjoying life
- an enhanced sense of spirituality or faith

The findings indicated that 65 percent of the participants were classified as high in cognitive processing and 40 percent were also able to find some meaning in the loss. Most of the men who engaged in cognitive processing were able to find meaning in the loss. Those finding meaning had better immune-system functioning and lower mortality during the follow-up period. Only the men who were able to find meaning maintained the immune status (e.g., CD4 T cells) over two to three years of follow-up. Those who could not find meaning in their bereavement, even if they were high in cognitive processing, showed immune-system declines during this period. Amazingly, among the men who discovered meaning, only 18 percent died during the follow-up interval, while *approximately 50 percent of those who did not find meaning died during this time.* The findings were not accounted for by initial immune status, HIV symptoms, emotions such as depression or loneliness, or health behaviors.

Finding meaning leads to better adjustment among mothers of preterm infants. Earlier I described a study by Affleck and colleagues that assessed how 114 mothers were able to cope with having a preterm infant in intensive care.[6] Many of these mothers were able to find meaning and discover benefits from their ordeal. The researchers wondered whether these mothers adjusted better even long after the infants were discharged from the hospital. Assessments of emotional distress and other factors were gathered when the infant was discharged from the neonatal intensive-care unit (NICU) and again at six and eighteen months later. At discharge three-quarters of the women consciously sought meaning in the crisis, 42 percent were able to find some meaning, and 75 percent were able to identify some benefits. Women who became more emotionally distressed over the ensuing eighteen months after discharge tended to be those who failed to seek meaning in or find benefits from the crisis. What was even more fascinating is that a mother's discovery of benefits was helpful for the infants. Babies of mothers who were able to construe benefits from the crisis showed significantly more rapid mental development and other adaptive behaviors over eighteen months.

Why would a mother's ability to find meaning be associated with improved development in these preterm infants? According to Affleck, "One 'suspect' that can probably be ruled out is optimistic expectations regarding the infant's outcome. A plausible line of reasoning is that mothers who found benefits in the NICU crisis were more optimistic about their child's development. Their optimism, in turn, may have fueled greater attention to the infant, which contributed to better infant outcomes. But we discovered that even after taking into account mothers' expectations, benefit-finding predicted infants' developmental outcomes." Affleck's collaborator, Howard Tennen, said, "It is also reasonable to speculate that friends, neighbors, and family members may find it easier to support mothers who acknowledge benefits because they appear more 'upbeat.' This support from others, in turn, may provide an important psychological and tangible resource that allows these mothers to attend to their infants with greater vigor. Indeed, we found that mothers who reported benefits in the NICU experience were less socially isolated. Yet mothers' social isolation did *not* predict their infants' subsequent development. Overall we have few solid leads regarding the mechanisms that might have linked mothers' benefit-finding with their own superior emotional adjustment and with their infants' development eighteen months after they were discharged from the NICU."

The research summarized in this chapter demonstrates that finding meaning can indeed lead to improved psychological, emotional, and physical health. But perhaps the strongest suggestion of the health impact of finding meaning comes from the research on religious faith and health. As you will see in the next chapter, religious participation is unequivocally linked to health and longevity. Not surprisingly, it is also one of the most prevalent routes to growth and meaning for many of us.

Chapter 12

Faith, Meaning, and Longevity

One warm, early-spring Sunday morning in the heart of North Carolina, my father was where he had been nearly every Sunday for almost fifty years—in the pulpit of his church delivering a typically emotional and thought-provoking sermon to a full congregation. As copastor, my mother was in her usual seat behind him in the pulpit, leading the call and response so characteristic of many African American churches. Near the end of the sermon, as he and the congregation were at their spiritual and emotional peak, he turned to my mother, his back to the congregation, gripped her hands in his, and looked into her eyes for a few moments, communicating something unspoken. He turned again toward the pews and collapsed, the victim of a heart attack.

Each of us must face our own mortality and that of loved ones. I understood my father's heart attack intellectually, given his age and risk factors for heart disease. He and I had actually talked about the possibility a number of times. I relate this story not because of what happened at the church that Sunday but because of what happened afterward. My father did not die that Sunday. In fact, he lived for almost two years—but in a completely vegetative, comatose state.

The last thing I ever imagined was that he would be in an enduring coma, and coming to terms with this was one of my most difficult experiences. I just somehow could not comprehend, could not fully assimilate the ongoing state he was in, somewhere between life and death. Here was this strong, 230-pound extreme extrovert, who valued the life of the mind and intense verbal exchanges, who was always in control of every situation, who was now silenced, motionless, bedridden, and, according to his EEGs, without consciousness. Through his eighteen-month ordeal, I struggled daily, even hourly, with its meaning. I constantly posed the *why* question to myself. "Why did this happen to a man who had dedicated his life to uplifting others?" Although I was well aware that "it rains on the just and unjust," making sense of his condition was hard. My father's coma was simultaneously an emotional and an existential challenge for me.

At times when my emotions reached a nadir, my mother's words often helped lead me toward meaning. During this time she often said to me, "We have to look beyond what we see in the physical world and see with the eyes of faith, with the eyes of the spirit. Remember that the victory has already been won." By "victory" she meant that, through faith, we are always able to conquer adversity in the physical world. My mother's worldview was explicitly spiritual and allowed her to cope with, understand, and find meaning in just about anything life threw her way. In fact, as a minister herself, she would probably say that her job description was largely helping people find meaning in life's tribulations. Sometimes it seemed as if our home telephone were an emergency hot line. Calls would come in at all hours from church members in varying degrees of anguish—bereaved widows, spouses whose partners were fighting addictions, parents of troubled teens, the elderly wishing to escape loneliness and isolation, or former church members who had lost their way. At times her advice needed to be directive—actions were necessary for the caller to overcome the dilemma. But more often than not, what she dispensed was a spiritual perspective. With a mixture of love and compassion, void of condemnation, condescension, or judgment, she helped the callers recalibrate their spiritual compasses and use their religious orientation to draw comfort, strength, and, on occasion, answers.

With my mother's spiritual perspective as a model, I began to rely more on my own religious and spiritual orientation—my spiritual schema. My spiritual perspective became stronger during the period of my father's coma, and it became the only thing that helped me to come to terms with his ordeal.

Spiritual matters are not typically discussed—and are oftentimes frowned upon—in scientific circles. In psychology the historical zeitgeist has been a kind of neutrality on the topic of religion and spirituality.[1]

Sigmund Freud wrote that religion was "the universal obsessional neurosis of humanity."[2] Freud was partly right. Religious practices and spirituality are indeed universal and widespread. Recent Gallup polls have indicated that 96 percent of Americans believe in God (94 percent of younger and 97 percent of older adults). Over 90 percent of American adults are affiliated with a formal religious tradition, with 46 percent reportedly attending weekly or almost-weekly worship services. Approximately 67 percent are members of a local religious body, and over 60 percent feel that religion is "very important" in their lives.[3]

Freud notwithstanding, a religious or spiritual orientation may have a value that goes beyond what we have considered. Research on the health effects of religion and spirituality has slowly accumulated over the last two decades. Because of its promise as a field of scientific inquiry, I formed a committee on religion and health shortly before I left NIH, with representatives from each of the then twenty-four NIH institutes (which focus on areas as distinct as cancer, heart disease, alcoholism, and mental health). The committee's goal was to reach consensus on the status of the science of religion, spirituality, and health. Were the studies well designed? Were the findings scientifically and medically significant? Was this field of study one that NIH should continue to pursue? The committee brought together the top scientists in this field from around the country to rigorously evaluate the quality of evidence from current research. We were not the first to examine this research closely, but it was the first time the NIH did so across all its units. Having NIH validate a field as legitimate is tantamount to giving it the scientific Good Housekeeping Seal of Approval. Therefore the committee would provide an unprecedented evaluation of this research.[4]

Of all the research reviewed, perhaps no scientist was associated with more studies than Dr. Harold Koenig of the Duke University School of Medicine. I met Harold years ago, when I was a professor at Duke and he was a research fellow just beginning his career. Koenig told me that although he was raised in a religious home, religion really was not a big part of his life during college and medical school. Regardless, religion still provided him with some degree of help when he was coping with various life trials. During his psychiatry residency his interest in religion as a scientist was awakened. He explained to me, "I began to ask patients what they did to help them cope . . . with their medical illnesses, and I was struck by the fact that so many of them mentioned religion. It was about that same time that I found that religion was a source of comfort for myself, and I just thought, 'Well, I'm not the only person who has these experiences.' There just seemed to be a lot of people who were using religion to help them cope with their illnesses and, in fact, find meaning in the illness."[5]

Over the years Koenig has become one of the true stars in research on religion and health. His recent book, *The Healing Power of Faith*, provides an accessible yet scientifically accurate account of what has been learned in this field and is congruent with the NIH committee's findings. Here is what we know about faith and health:

Religious participation predicts longevity. One of the most fascinating areas of research on religion concerns its influence on longevity. In 2000 Dr. Michael McCullough and colleagues published a comprehensive review of the science of religion and mortality.[6] They used a sophisticated statistical approach called a meta-analysis, which is a mathematical method for drawing conclusions from a large number of studies on a single topic. Their analysis was based on twenty-nine studies, which collectively involved more than 125,000 participants. Health outcomes of the participants in these studies were tracked for various periods of time, from as few as two months in a study of terminally ill patients to more than twenty years in a study of a healthy population. To assess religious participation, subjects were asked questions about the frequency of attendance at religious services, how religious they felt they were, the role of faith in their lives, or their participation in private religion activities (e.g., praying, watching religious TV programs).

Summarizing many studies, the researchers concluded that people who scored higher on measures of religious involvement were nearly *30 percent less likely to have died during the period represented by the study* than were those who scored lower on religion measures. The increased longevity among the more religiously involved was true even after statistically accounting for physical and mental health, gender, race, health behavior, and social support.

One of the studies included in the McCullough review was a project by Dr. William Strawbridge and colleagues.[7] Strawbridge analyzed the relationship between church attendance and death over a twenty-eight-year period in a study of 5,200 residents of Alameda County, California. Frequent church attendance was defined as attending services once a week or more; infrequent attendance was defined as attending once or twice a month or less. Over this time church attendance (or lack thereof) was fairly consistent—people who were frequent or infrequent church attendees at the start of the study in 1965 remained so at the end of the study in 1994.

The risk of dying over nearly three decades was 36 percent lower for frequent church attendees than for infrequent attendees. The significant difference between the two groups remained after adjusting statistically for age, health status, social connections, and health practices. Some critics of research on religion and health have argued that studies do not take into account the fact that people who are already physically impaired would likely attend services less often and that the high death rate for these infrequent attendees is due to preexisting illness, not lack of church attendance. Interestingly, not only did Strawbridge statistically account for this possibility, he found that the frequent church attendees were somewhat *more impaired* than their infrequent counterparts, yet they lived longer. In fact, a recent major eight-year study involving a nationally representative sample of more than 20,000 people found a dose-response relationship between religious participation and health. During the eight years, compared to people who attended church more than once a week, *those who attended just once a week had a 15 percent higher rate of death, those who attended less than once a week had a 31 percent higher rate of death, and those who never attended had an 87 percent higher death rate.* The association between religious attendance and death remained significant even after accounting statistically for health status, economic factors, age, race, gender, and region of the country.

This kind of careful attention to methodological rigor is characteristic of many studies in this field, whose findings are summarized below. Measures in these studies include such factors as the frequency of church attendance, perceived closeness to God, denominational affiliation, the use of private prayer, depth of religious faith, frequency of Bible study, and the use of religion to cope with stress.

Religious participation predicts illness. A number of studies have linked measures of religious beliefs and practices to specific illnesses, such as high blood pressure, coronary heart disease, and disability.[8] For example, one study found that people who were very active in religious activities were 40 percent less likely to have hypertension than were people who participated infrequently in religious activities.[9] Higher religious participation is related to the development of fewer illnesses over time, even among people who are initially healthy.

Religious participation predicts emotional well-being. People who are regular churchgoers report more satisfaction with their lives, more emotional well-being, and lower levels of depression than do others.[10]

Religious participation predicts immune status. In studies at the University of Miami involving 106 HIV-positive gay men[11] and at Duke University involving over 1,700 elderly persons,[12] religious participation and beliefs have been associated with enhanced immune-system functioning.

Religious participation predicts use of hospital services. Research at Duke and other universities has found that the higher the level of religious participation, the lower the likelihood of being admitted to the hospital. The more religiously oriented patients spend fewer days in the hospital than do the less religiously involved.[13]

Religious participation predicts positive health behaviors and perceived social support. Active religious participation or strong religious beliefs are linked to low rates of cigarette smoking, high rates of smoking cessation among smokers, less alcohol abuse, high levels of physical activity, less use of illicit drugs, and high levels of perceived social support.[14]

THE POWER OF FAITH—WHAT IS THE MEANING?
IS IT IN THE MEANING?

Maya Angelou, who was asked by President Clinton to create and deliver a poem at his 1993 presidential inauguration, is familiar as a performance artist and a celebrity. We know either of her poetry, her books, her plays, her political activism, her career as an actress, her beautiful speaking and singing voice—or her friendship with Oprah. What many of us do not know is that Maya Angelou's life is fueled by faith. She explained to me, "It has been my fortune to step out on the word of God and find it steady."[15]

As a child Angelou saw faith actualized in her grandmother, and she relies on it constantly. She admits that faith is for many, herself included, a kind of elusive, paradoxical commodity. She described it to me as "a steel wall and such a will-o'-the-wisp." She reminded me of a story from the New Testament in which a man who encounters Jesus is quoted as saying, "Father, I believe. Forgive my disbelief." "What a relief," she said, that someone else had felt like that. Angelou added, with her characteristic humor, "My faith is absolute, until I question it."

She had no question or doubt in 1983 when she received a call from the hospital where her son, Guy, was having his third spinal operation. His surgery had been very long but successful, so she was astounded to hear the doctor say, "We are losing your son." She went immediately to Guy, but before entering the room, and with the conviction and courage borne of her grandmother, she quietly beseeched him, "Hold on to your life!"

Guy indeed escaped the threat of death, but the doctors told Angelou that he would be paralyzed. Despondent, Guy said to her, "I know you love me, and I know I am your only child, but I have to ask you to do something no one should ask another. If there is no recovery, please pull the plug." His words stunned her. Fear, and faith, caused her to unleash the power of her voice as she sought to move heaven and earth on his behalf. She explains, "I can't even tell you how loud my voice is. It's window breaking, and I let it go!" Angelou vociferously cried out, commanding her son and pleading with her

God, "RECOVERY! I see you walking! I see you swimming! I see you playing basketball!"

She left her son's bedside for Mission Dolores in San Francisco, for her a place of spiritual sustenance. When she returned, the doctors reported no progress and discouraged what they called "a mother's hopes" that Guy would walk out of the hospital. They explained to her how fragile the spinal cord is and how a blood clot had violated Guy's spine for eleven hours. As Angelou continued telling me the story, I thought about how much she reminded me of my own mother, who was able to see events through the eyes of faith, not only through the eyes of medicine. Angelou said that she told the doctor, "I am not asking you, I am *telling* you. I went somewhere far beyond any place you can dream of! Thank God, my son *will* walk out of this hospital."

Three days later a nurse summoned Angelou to Guy's room and lifted the sheet from his feet—Guy moved his toes.

I share this story from Maya Angelou because it demonstrates her unwavering faith—a faith that for her and many others comes dramatically to the fore when circumstances seem most dire. Reliance on religious faith to cope with stress may be one reason it has such a powerful and pervasive effect on health and longevity. But there may be other reasons as well. Religion is associated with known health-protective factors, such as higher levels of social support, moderate alcohol intake, less cigarette smoking, more physical activity, remaining married, and participation in meditative activities. In other words, people who value religion also take better care of themselves. But many of the studies documenting the health effects of religion indicate that religion provides something beneficial to health that is over and above lifestyle and relationship factors. But what is that something?

One possible answer is that religion and spirituality provide people with frameworks or schemas for confronting life's difficulties. Religion, you might say, is a path to meaning. That is, it helps the individual make sense of tumult or find benefits in it. In his book *The Psychology of Religion and Coping*,[16] Dr. Kenneth Pargament of Bowling Green State University notes that both historically and scientifically it is clear that a central function of religion for many of us is that it provides a framework for life, a way of comprehending the world. Other researchers, such as Dr. Daniel McIntosh of the University of Denver, believe that religion's ability to help people cope with crises derives from its schema-like functions. Religion helps people more rapidly assimilate or accommodate adverse experiences.[17]

The religious implications for survivors of the Holocaust might seem beyond comprehension. They had been targeted for annihilation because of their religion. Yet most survivors relied on their religious faith and their trust in God to assimilate what had happened and to find meaning. Helmreich reports findings in *Against All Odds* that "seven out of ten survivors did not change their religious behavior as a result of the Holocaust." Some became less observant immediately after the war, he points out, but almost half of those who did reclaimed their observant behaviors. Even those who harbored doubts about what happened continued to believe in God. The fact that they survived the Holocaust was reason enough for some survivors to justify their continued faith. Some felt that God had saved them for a reason, which might have precipitated their finding of meaning. Religious practices themselves can create meaning and become a way for many survivors to honor family who died and maintain an unbreakable link to their past. Helmreich reports one survivor as saying,

> I believe that it was my destiny to survive, to come back from the ashes and to be the link from the past that would begin life again and pass it on to future generations. My goal was to remain faithful to the religion into which I was born and to my upbringing. This upbringing gave me the strength to survive the war. When I live this way, I know I am living the kind of life my departed parents would have wanted me to live.[18]

Religion is used by many as a way of interpreting both positive and negative events so that they are more easily assimilated. Another example of how religion operates to provide meaning comes from the research on the fear of death. People who are deeply religious experience significantly less anxiety about death than do others. In one study only 10 percent of the participants who used religious beliefs to cope with stress expressed anxiety about death, but anxiety occurred for 25 percent of those who did not.[19] For many religious people, then, death is not a schema-exploding notion. Many find the idea of death congruent with their assumptive worlds.

Has science been able to document the idea that religion's health-enhancing effects are due to its ability to provide meaning? Research on this question is far from definitive, but there is some circumstantial evidence about religion's ability to provide meaning. It comes from research on the use of religion at times of stress.

Lean on Me

In my parents' church the choir used to sing "Leaning on the Everlasting Arms," a line of which is "leaning, leaning, safe and secure from all alarms, leaning, leaning; leaning on the everlasting arms."

This hymn captures the essence of a burgeoning subfield within the religion/health domain known as *religious coping*. Religious coping goes beyond the mere act of church attendance or saying that you value religion. It is the active use of religious activities to cope with stressful life events. This may involve seeking strength and comfort from God, prayer, confessing sins, soliciting support from clergy or church members, or using religion to gain a sense of coherence or meaning. People who use religious coping activate these or other religion-based processes to carry them through adversity. According to Harold Koenig,

> The basic elements that give meaning and purpose to the religious person's life are not easily threatened, even by dramatic life changes such as financial reversals, serious illnesses afflicting oneself or loved ones, or the death of a child or spouse. Because the annoyance and hectic pace of daily life do not threaten religious people's underlying values, their perceived stress levels are lower than those of the less religious. Faith helps mitigate the initial discouragement and hopelessness provoked by negative experience, which can steadily accumulate to debilitate people.[20]

Angelou told me that she has countless examples of how her faith has helped her overcome, understand, and generally cope with life's travails, because she uses faith and her belief all the time. With the dramatic characterization for which she is famous, Angelou described for me the persistence of her faith. "When bad things happen to me, before I speak to anybody, I speak to God. I say, '*Here it is.* I don't know how to get over this one, but here *it* is. It is very heavy, very painful, and burdensome. Here *it* is. Whatever I did to bring it about, I am sorry. *Here it is.*' Now, if it tried to follow me, I would go back and lay it down. If it wakes me up at night, I say to it, 'Wait a minute—hold on.' Then I say to God, '*Here it is.* I don't know what to do with it.' And, amazingly, it keeps getting smaller and smaller and smaller. Yes," she affirms with resolve, "I take my burdens to the Lord, and I leave them there."

Angelou is not alone. Religious coping is common. Surveys indicate that when people are asked, "What enables you to cope with the difficult or stressful events in your life?" a sizable percentage mention a topic related to religion, such as God, prayer, attending church, or faith.[21] If attending religious services or believing in a higher power is the structure of religion, then religious coping might be thought of as part of the function of religion. That is, it is one way people actually use religion in their lives.

Does it help? It certainly seems so. Research has demonstrated that the use of religious coping ameliorates some of the untoward effects of stress. Psychologist Dr. Kenneth Pargament, one of the top scholars on religion and health, told me, "Religious coping can play a number of valuable roles in the lives of people. It can help people hold on to a sense of meaning in the face of events that may seem to make little sense at all. It can instill a sense of connection to something greater than oneself in the midst of situations that tend to separate us from each other. It can support and strengthen us when we are feeling at our weakest. And it can help us transform our most fundamental values and visions in life when old sources of significance are lost or no longer viable. Part of the power of religious coping lies in the fact that it can help meet the diverse needs of people facing very different problems in very different environments."

To determine whether religious coping lessens the impact of stress, researchers typically ask study participants a series of questions about current or previous life stressors and the degree to which the participants use or have used religious faith to counteract those stressors. For example, they may be asked the degree to which statements like the following are true for them:

- My religious faith helps me cope during times of difficulty.[22]
- When dealing with difficult times in life, I get much personal strength and support from God.[23]
- Prayer helps me cope with the difficulties and stress in my life.[24]
- I tried to find a lesson from God in the stressful event.[25]
- I trusted in God to handle the [negative] situation.[26]

Research indicates that people who report a greater reliance on religious coping during times of stress have an emotional, and potentially a physical, health advantage over those who do not. Here are four examples of what has been found:

Religious coping is associated with lower depression scores. In a study at Duke Medical Center, Koenig and colleagues studied 850 older patients admitted to the hospital. Those who relied more on religious-based coping strategies to handle emotional stress were significantly less likely than others to have depressive symptoms. The association between religious coping and depression persisted after accounting statistically for race, sex, age, social support, and health variables.[27]

Religious coping predicts better adjustment to transplant surgery. At the University of Minnesota, researchers assessed religious coping, emotional distress, and life satisfaction in kidney-transplant patients and in their significant others. At both three months and twelve months after surgery, patients and significant others reporting greater use of religious coping had lower distress and more life satisfaction than did others.[28]

Religious coping lowers ambulatory blood pressure. Dr. Andrew Sherwood and colleagues examined the effects of religious coping on blood pressure in 155 healthy men and women (78 African Americans and 77 whites). Participants wore portable blood-pressure monitors for twenty-four hours on a typical workday. The monitors recorded and stored blood-pressure readings four times during the day and twice at night. Higher levels of religious coping were associated with significantly lower blood pressure, especially for the African American participants.[29] The greater benefit of religious coping for the African American participants could have been a result of the fact that religious coping has consistently been found to play a larger role in the lives of African Americans than in those of whites. In this study and numerous others, African Americans scored higher on measures of religious coping than whites did. African Americans, more so than whites, use religion as a stress-coping device that is more strongly tied to well-being.[30]

Religious coping predicts mortality. Very few studies examine religious coping and mortality, but Dr. Neal Krause of the University of Michigan conducted an interesting one, which assessed religious coping in more than 800 older adults.[31] Participants were asked about stressors in their lives associated with various roles, such as those of parent, spouse, homemaker, and volunteer. They were also asked about the degree to which they relied on religious or spiritual orientation to help cope with such stressors. Initial assessments were completed in 1993, and follow-up assessments to determine who had died were conducted approximately four years later. Krause found that the use of religious coping acted as a buffer against the effects of stress on mortality among certain participants. Those who used religious coping in response to stressors in roles they valued highly were less likely to die during the four years than were those who did not. This was especially true for persons with less formal education, for whom, many studies have found, religion tends to play a large role.[32] Pargament also noted from his research that the benefits of religion likely accrue for those most deeply involved in it.[33]

In another study of religious coping and mortality, researchers at the University of Texas Medical Branch in Galveston studied more than 200 patients for six months after elective open-heart surgery to determine what factors predicted survival. Six months after surgery three biomedical variables were strongly related to survival: (1) whether patients had previous cardiac surgery, (2) whether patients had im-

pairments in basic daily activities (e.g., getting dressed) before the surgery, and (3) whether patients were older. After accounting statistically for these factors, it was discovered that patients were at greater
risk for death if they reported gaining no strength or comfort from religion or lacked participation in social groups. Moreover, as seen in
Figure 12, the patients at the highest risk for death after surgery were
those who lacked both religious strength and comfort and group participation.[34]

How Religion Enhances Longevity: Is It the Meaning?

Andrew J. Young, a top aide of the Reverend Dr. Martin Luther King
Jr. during the civil rights movement, is himself an ordained minister.
He also served three terms in the U.S. House of Representatives, was
ambassador to the United Nations under President Carter, served two
terms as mayor of Atlanta, and cochaired the Centennial Olympic

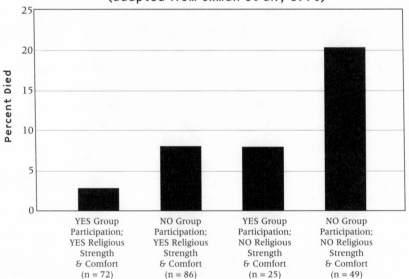

Figure 12: Mortality Associated with Social
and Religious Involvement
(adapted from Oxman et al., 1995)

Participation in Social Groups (Yes or No) and
Gaining Strength and Comfort from Religion (Yes or No)

Games in Atlanta in 1996. Not bad for a man who, after college, had no plan but left his fate in the hands of God. First of all, you have to understand that Young hailed from what he describes as a "very good, strong, and religious family." Still, it was his own epiphany that he says pulled everything else into line. Young explained to me that he had just graduated from Howard University but felt lost and that his life was meaningless. His father wanted him to become a dentist, a dream Young did not share, but he had no other viable goal. During a visit to North Carolina, he pushed himself to the brink of physical exhaustion running up a mountain trying to release his frustration at being what he described as "naïve about life, despite a college degree." A revelation was waiting for him at the top of that mountain. He said, "It was almost in a moment like a flash of insight, looking out on those North Carolina fields and hills and the sky—everything seemed so orderly and purposeful. It suddenly hit me that there must be a purpose in life for me." Young said that a burden had been lifted, and in the next few months his life took shape as he believed God intended.

He accompanied his pastor to a religious youth conference in Texas with the intention of spending time with his college roommate, who lived in San Antonio, about 150 miles from the conference site. When they arrived, his pastor encouraged him to remain at the conference for moral support. They had not seen any other black people since they had entered the Texas panhandle from Dallas, and according to Young, his pastor told him, "You're not going to leave me here by myself, and I know you don't want to ride down these roads by yourself." Young ended up staying, and he began a journey that culminated in his going to seminary in Connecticut.

He participated in the conference and was impressed by the commitment he observed in many young white people who were having their first interracial interactions. He said, "It was the first time I had been around white people where their religious faith made a difference in their conduct, and that was impressive." He ended up as a volunteer for the National Council of Churches, which had sponsored the youth conference, and he was assigned to Connecticut. When he arrived, no housing arrangements had been made, so he was placed on the campus of the Hartford Theological Seminary. Since his volunteer work took up his afternoons and evenings, he requested permission to audit a couple of classes; school officials suggested that he might as well take three to qualify for a scholarship. He enrolled and did well while the school waited for his transcript from Howard, which took about two or three months to arrive. Good thing, because the school informed him that had they seen his grades, they would not have been able to admit him. Young reflects, "So through all of this, I felt God had a plan for my life, and it was working out, and all I had to do was go along."

Young believes that God, with just as firm a hand, continued to direct his life, sending him to pastor a church in Alabama and placing him in the home of his future wife. She wasn't even there, but he knew before they met that he would marry her. He tells the story better than I would: "When I got to Alabama, in the first house that I visited, I saw a Revised Standard Version of the Bible on the table. It had been underlined, and the name on it was Jean Childs. This was 1951, and this version didn't come out until 1950, and so what was a black woman in the country doing with a Bible that had just been published, and it was already underlined, and in a lot of my favorite chapters. Her mother had told me the book belonged to her daughter, who was away at Manchester College in Indiana. I had been to Manchester the summer before for a camp on my way to seminary. It's a college that advocates nonviolence, and it's where I started reading Gandhi. So to have some young black woman who reads the Bible and who probably understood nonviolence was more than I could have imagined. Then I looked on the wall in the living room, and there was a senior lifesaving certificate, and I had been on Howard's swimming team. I didn't know any women who could swim. So I just decided that the Lord must have sent me there for a wife. Sure enough, two years later we got married. There was never any doubt in my mind that she was the woman I was supposed to marry, even before I met her, and that's the way God has planned for me."[35]

For Young and others religion is one of the chief frameworks to bring meaning and coherence to their lives and to help them understand events, both negative and positive. And, as summarized earlier, religion is strongly associated with better emotional and physical health and longevity. Religious-based coping in the presence of stress protects against depression, improves life satisfaction, and may be associated with longevity.

But have we really answered the question of whether religion has its salutary affects *because* it provides meaning? This is a tough question, since religious beliefs and practices do more than provide meaning. They can enhance other factors that improve health, such as social support and beneficial health behaviors. But some of the most rigorous studies in the field have taken such factors into account and yet still detect the link between religion and health.

One fascinating study takes us even closer to understanding the explicit role of meaning in the religion/health connection. Dr. Daniel McIntosh and his colleagues wanted to determine the role of religion in helping people cope with one of life's most severe traumas: the death of a child.[36] The participants in their study were parents of babies who had recently died of sudden infant death syndrome (SIDS), a term used to describe the unexpected and, until recently, largely unexplained death of an apparently healthy infant.[37] McIntosh chose this population of parents because the loss is so shocking that no psychological preparation is possible. Parents of the deceased children were asked, among other things, about their level of religious participation and the importance of religion in their lives. The researchers asked questions to determine if such religious beliefs and practices helped the parents adjust to their loss. Significantly, they wanted to ascertain *how* religion helped, so they also explored whether religion enhanced adjustment by providing parents with more social support, by providing them with meaning, or by helping them *cognitively process* the loss. In this case researchers defined cognitive processing as having recurring thoughts and images about the child. People who had more thoughts, memories, and mental pictures about the child and who purposely engaged in thinking or talking about the baby were said to be high in cognitive processing of the death. Although difficult to endure, cognitive processing is ultimately beneficial. It helps a person assimilate, or make accommodations for, traumatic events.

The researchers thought that religion might speed up this cognitive processing because it could provide parents a schema for thinking about and understanding death (e.g., that their child lives on in

heaven, that they will see their child again). The question then became, Does religion help a parent cope with the sudden loss of a child
through social support, cognitive processing, or finding meaning?

To answer this question McIntosh and his colleagues looked at
two aspects of religion: religious participation (church attendance)
and the importance of religion to the participants. The researchers believed that higher levels of religious participation might help parents
cope with the death of their child by exposing them more frequently
to a supportive church community. People who say that religion is
very important to them may cope better because they develop a
strong religion-based schema or worldview that would help them
both to find meaning in their loss and to cognitively process it. Figure
13 shows the ways in which religious participation and the impor-

**Figure 13: Ways by Which Religion Participation
and Viewing Religion as Important
May Affect Adjustment (adapted from
McIntosh et al., 1993)**

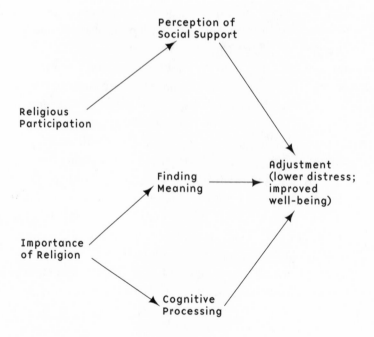

tance of religion would ultimately affect adjusting to loss, in terms of the amount of psychological distress and emotional well-being experienced by the parents.

In the study McIntosh interviewed parents on two occasions—at three weeks and again at eighteen months after the child's death. The findings supported the authors' hypotheses about how religion might affect distress and well-being after a major loss. Greater religious participation was strongly related to increased social support. Participants who said that religion was important to them were more likely to find meaning in the loss, and they experienced better cognitive processing.

Religious participation was also related to the *ability* to find meaning, suggesting that the church community aided the parents in understanding their loss. In turn, these three factors—social support, finding meaning, and cognitive processing—were related to higher levels of well-being and lower levels of distress at three weeks and eighteen months after the death of the child. This is one of the few studies to explicitly examine and link the process of finding meaning with the positive emotional effects of religious involvement.

According to McIntosh, religion's health-enhancing effects may indeed be in its ability to bring meaning to experience, "This is more than an intellectual exercise. How we think about, categorize, and understand events—what we believe causes them and what we believe the long-term outcomes might be—influence our behavior. How we emotionally and behaviorally respond to any event is influenced by our understanding of the event. For some events more than others, and for some people more than others, religion provides the framework through which the event is understood. Because everyone's beliefs are influenced by people in their social network, participation in a religious community provides a constellation of people who help develop our beliefs and who can contribute understandings of ongoing events in our lives."

Other Routes to Meaning

I've written a lot here about religion as a way of finding meaning, largely because that's where the majority of the science is right now. But certainly religion and spirituality are not the only ways to find meaning. Harold Koenig related his thoughts to me on this topic: "Nonreligious people find meaning in many other things. They find meaning in terms of what they contribute to society, what they give to their families and their children, and what they accomplish in life. All those things provide the person's life with meaning. Howard Hughes may have found meaning in accumulating millions and millions of dollars. Others find meaning in their grandkids, their families, or their relation to others. I can't imagine getting any closer to finding meaning than that."

There are countless ways to move toward finding meaning after adversity that are not religious in nature. Many people turn their tragedy into activism. Jannette Fennell is an example. In October 1995, a carjacker left Fennell and her husband for dead in the trunk of their car.[38] Having trouble breathing, she ripped away at the interior of the trunk until she found a way out. She later translated her terror into a crusade for emergency trunk releases. Fennell learned that 260 individuals had died since 1970 because they became trapped in a trunk—this included about 40 children under the age of fourteen. Most were trapped accidentally and died from heatstroke and suffocation. The force of her fury changed the automobile industry. Ford Motor Co. responded by installing handles that glow in the dark and can be easily opened by children. General Motors designed a system that senses body movement and temperature, which cause the trunk to open automatically.

Alan Chick, a police officer in Fort Worth, Texas, was helping a motorist when a drunk driver with two dozen previous arrests mowed him down. His wife, Lisa, was shattered, but she had a two-year-old son and an eight-year-old daughter who desperately needed her. Lisa Chick pulled her family together with the help of grief counseling and reached out with volunteer work. Out of her own pain she has found the strength to advise police officers on how to plan for their families if they, like her husband, are killed in the line of duty.[39]

Terry Fox was eighteen years old when bone cancer forced doctors to amputate his right leg six inches above the knee. The night before the surgery, Fox read about an amputee who had completed the New York Marathon. After having seen the suffering of other cancer patients and being inspired by the article, he decided to run across his homeland, Canada, with the aid of a prosthesis, to raise money for cancer research. Fox began his trek, the Marathon of Hope, on April 12, 1980, running more than twenty miles a day for nearly five months before the cancer stopped him for good. He died at the age of twenty-two, but not before turning his tragedy into something that changed the world. Fox was told before he died that the Marathon of Hope would be an annual event. The next year, 300,000 people ran in Canada and raised $3.5 million. By 2000, 1.5 million people were participating, including people from fifty-five countries outside Canada, and $21.7 million was raised. Close to $300 million has been raised for cancer research in the name of Terry Fox. He not only found meaning for himself but inspired others to action as well.[40]

Daniel and Joni Evans lived every parent's nightmare when a police officer knocked on their door a little after midnight to tell them that their daughter, Rebekkah, had been killed in a car accident. She'd lost control on a rain-slicked Georgia highway, crossed the median, and hit a Jeep. She, her best friend, and two other teenagers were killed. Rebekkah's parents will live with haunting memories, but Daniel Evans transformed his grief into a way to honor his daughter. He became a national spokesman for the Safe America Institute, which is dedicated to making America safe for young people. "It's a way to give her life and the accident some purpose and meaning," Evans said. "While it was a personal tragedy for me to lose Rebekkah, it would be an even greater tragedy if I didn't turn her loss into something positive."[41]

Tim Streett saw his father, Alan, an army major, paratrooper, and chaplain, gunned down in their driveway for a dollar. Tim and his father were shoveling snow when two young men approached and demanded money. His father was shot before he could respond. Tim expected to be next, but the killers grabbed his wallet, which held a single dollar, and ran. Tim watched his father die. Three men were arrested in a few weeks, but Streett's life unraveled; he drank, used drugs, and drifted from job to job. Almost ten years after his father's death, he reached an epiphany that led him to follow his father's footsteps into the ministry and devote his life to saving inner-city kids, like the ones who'd killed his father. He even moved his family to one of the roughest neighborhoods in Indianapolis to live near the youth who were to be the beneficiaries of his new commitment. Streett later decided to reach out to his father's killers and sent each a letter. Don Cox, who had driven the getaway car that was parked around the corner, was the only one to respond. Although Cox had not been at the scene, he had accepted personal responsibility for his involvement and was working toward rehabilitation. The two men met, and Streett embraced Cox in his ministry, visiting him often and extending forgiveness for his role in the murder. One of the most common themes in religiosity is forgiveness, which often contributes to making peace with and finding meaning in tragedy. Streett began to think that Cox had been severely punished, given that Cox wasn't directly involved in his father's murder. He successfully appealed to the judicial system to reduce Cox's sentence to twenty-three years and release him. Cox became an auto mechanic, a skill he learned in prison.[42]

Zora Kramer Brown has been a breast-cancer survivor for almost twenty years. When diagnosed, she wasn't traumatized and never needed to ask "Why me?" Along with a strong faith in God and an appreciation for the power of prayer, breast cancer was also part of Brown's inheritance. You see, her great-grandmother, grandmother, mother, and two of her three sisters had also been treated for breast cancer. Her grandmother lived to be ninety-four years old, and her mother is a forty-year survivor. Brown hails from a long line of women survivors, and they put their trust in the "grace of God." Still, when her sister Belva was diagnosed with terminal breast cancer, Brown was almost inconsolable. She wrote, "I could not understand why this was happening to her, and I cried so much that I almost choked on my own breath." She says she fell to her knees and asked God to help her through and to guide her in finding ways to help Belva. In return she promised to use all her resources and time for whatever work God called her to. Belva had encouraged Brown to become involved in breast-cancer awareness, and a kitchen summit of friends evolved into the Breast Cancer Resource Committee, an organization that provides support and education on breast cancer for African-American women. Brown says, "Through something as devastating as breast cancer, I found a way to help people, to encourage them, and that was and remains a tremendously gratifying work."[43]

CONCLUSION TO PART V

Does adversity have an upside? When adversity rears its unseemly head, the last thing you want to hear is "Hey, this is a good thing." But sometimes, perhaps once you've weathered the initial onslaught, closer examination of the full consequences might turn up some unanticipated benefits.

My experience with my mother's cancer is a good example. Despite my mother's unique leadership role in our church, she had, in many respects, fairly traditional views on her role at home. In our home my father was the dominant figure. He set the agenda for my dialogues with my parents. My mother was always present but always deferred to my father, and almost never offered opinions that contradicted his—when she did, it was usually with his encouragement to do so. So growing up, I was much more versed in my father's feelings, beliefs, and perspectives than I was in hers. Her influence was substantial, but it was instilled in less overt, less direct ways and more by example.

All this changed during my mother's bout with cancer, which occurred after my father's death. Our relationship grew in ways that I never imagined it would. I spent more concentrated time with her than I had in my entire adult life. We talked long and deeply, and I learned things about her and her life that I never knew. One of our favorite things to do was to take long drives in the country, during which I would ask her questions about her childhood and she would, at my request, record her answers into a tape recorder. I was able to say things to her, and she to me, that we were never able—or never took the opportunity—to say before. At times when her sickness prevented our drives and made it difficult for her to talk, it was a joy for me simply to be near her, holding her hand, saying nothing.

My mother's cancer could by no means be described as a "pleasant" experience. If I could, I would of course choose that she not have had to cope with it. But in looking back I view my time with her during her illness as among the most valued of my life.

The take-home message from this section of the book is by no means that adversity is good. It's not that when painful events occur we should immediately search for the upside or shut off the emotional turmoil. Stressful circumstances do not always lead to benefits. Bereavement, injuries, sickness, and other losses or setbacks are awful experiences. To suggest otherwise would be disingenuous. Even the act of finding meaning or benefits does not mean that the negative sequelae of trauma are obliterated. Trauma victims, even after finding meaning, may still struggle with the many consequences of stress, including depression, anxiety, and anger over their condition. They may still suffer physical pain and discomfort and may still have difficulties making long-term plans. Finding meaning or benefits may coexist with these aversive outcomes, but this coexistence can act as a kind of existential balm, making the experience more coherent and comprehensible, more easily assimilated, allowing us to draw back from it a bit, to put the experience in a broader context. To hold the experience up to the light to examine *all* its true implications, not just the most negative ones. A more complete examination of events might indeed uncover a surprising upside—renewed relationships, better appreciation of life, untapped abilities, strengthened faith. Such a process is intrinsically valuable and may in some cases serve as a counterweight to the extreme negative emotions that often accompany trauma and adversity.

Another message from the last three chapters can be posed as a question: Why do we so often wait for adversity to strike before we begin to appreciate the good in life? I'm fascinated by the fact that the overwhelming number of studies on finding meaning address its discovery following adversity. It is almost as if adversity were an existential wake-up call, telling us that life is about much more than work, bills, worry, and material possessions. Most of us *know* that life is more than these things, but we're so consumed with running the race that we never catch the scenery. Adversity does not create the good in life or transform our priorities. But it does cause us to recognize and value these things. It refocuses our attention, draws us back to the emotional and psychological place we might have intended to be all along but missed—or simply never thought was available to us. So the question is, Why wait for adversity in order to find a larger meaning and purpose?

PART VI

CONNECTIONS—TO EMOTIONS, TO THE FUTURE

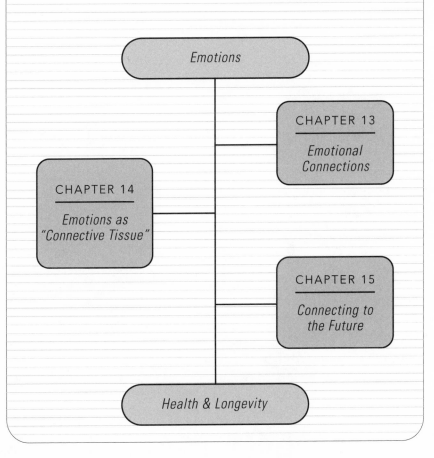

Emotions

CHAPTER 13

Emotional Connections

CHAPTER 14

Emotions as "Connective Tissue"

CHAPTER 15

Connecting to the Future

Health & Longevity

The future is here. It's just not widely distributed yet.

—William Gibson

Coming Full Circle

The last part of this book brings me full circle in many respects. I begin it as I began the first part, with a reflection on my mother's character and nature. When I decided to write a book about a more comprehensive view of health, I wanted to include in it stories about a person whose life was an embodiment of health's many determinants. I even created a composite of what such a person would look like—psychologically, socially, spiritually, and emotionally. Then I gave up on the notion, thinking, "This is silly; no one is really like this, and if such a person existed, he or she would seem just too good to be true." Yet the composite had a vague familiarity, as if it *was* in fact someone I knew. It was then that I realized I was actually describing my mother. I was initially surprised by this revelation, not because I didn't think she was wonderful but because she seemed to so perfectly fit the "profile" of the person I was looking for. So much so that I had to check frequently with Elizabeth to make sure I wasn't simply expressing my affection for my mother by force-fitting her life into the dimensions of health I would be describing in the book. Elizabeth, who knew my mother quite well and who never fails to tell me the truth (even when it isn't what I want to hear), assured me that I was not simply being sentimental. If anything, Elizabeth agreed that many who knew my mother probably will say I downplayed her nature. Elizabeth's perspective, too, is that my mother's character was almost eerily exemplary of what the book was about.

My mother was the consummate optimist, always expecting the best and always explaining the past in ways that were uplifting. Although she didn't keep formal private journals, she would write deeply about experiences that had affected her emotionally in a spiral-bound notebook or on single pieces of paper. I have found some of these pieces of paper inside old Bibles of hers. She was also fortunate to be in a profession—the ministry—that sanctioned the public disclosure of emotionally trying events. My mother had a large social network, and she was both the giver and receiver of health-enhancing social support. Although money was never important to her, she valued education. Her position as a minister carried with it a good measure of social standing in our community. And when it came to finding meaning from adversity, she had to look no further than her religious and spiritual worldview.

In many respects using my mother as an exemplar in this book on emotional longevity personally brings me full circle, from the perspective I had of her when I was a child to seeing her in a completely new light from the vantage point of a health professional. Growing up, I never really thought in any detail about all the specific things that made her special or what made her life fulfilling. She was simply my mother, and when I was with her, I was home, wherever we were. Ironically, it was my leaving home to pursue a college education and my twenty years of work as a researcher, psychologist, and administrator in the health field that have allowed me to see her in a way I never had before. In graduate school in the 1970s, I learned about a new area of science and professional practice that was just emerging— behavioral medicine. As a field, behavioral medicine has the same premise as emotional longevity—that physical health is a function of the interconnections between psychosocial, behavioral, and biological processes. Behavioral medicine was one of the first explicitly biopsychosocial scientific movements. In 1979 I attended the first meeting of the Society of Behavioral Medicine (SBM), and it changed my career path forever. Previously my professional interests were in childhood mental illnesses, but after that SBM meeting I began working on non-drug treatments for chronic pain; behavioral causes of heart disease; and stress and hypertension. As a professor at Duke, I ran a cardiovascular psychophysiology laboratory and worked closely with colleagues from many medical and social-science disciplines. Although my work and that of others in behavioral medicine has been varied, it is ultimately about the same thing—connections between the dimensions that make us who we are, and our connections with the environmental contexts in which we live. When I was recruited to NIH in 1995 to head the newly created Office of Behavioral and Social Sciences Research, I was attracted to the position because it would allow me to be on the front lines of the movement to integrate this broader, expanded view of health into health science. In fact, Congress, in creating the office, actually mandated that the NIH integrate this perspective. During my short time at Harvard, my focus was on communicating to the widest possible audience how far science has come in understanding the critical importance of the nonbiological aspects of health, and using this information to aid in the elimination of the profound health disparities that exist in our society. And now, as chief executive officer of the American Psychological Association, one of my main objectives is to work to change our health-care system to accommodate the new knowledge we have of the connections between the mind, behavior, the social environment, and our health.

Some thirty years after leaving home to pursue an education and establish a professional life where the consuming passion has been exploring the many connections that determine our health, I have mentally returned home and have psychologically come full circle. My work has allowed me to see with clarity my mother's life and her nature in all its richness and harmony. My mother was indeed the embodiment of health, in all its dimensions.

These final chapters also bring us full circle to the last dimension of the new definition of health—the emotions. But these chapters— indeed, this entire book—are principally about connections. Thus far I've provided scientific justification for expanding our view of health, for recognizing the manifold connections between psychological, behavioral, and social phenomena with health outcomes. This last section is about three types of connections. The first is the connection between our emotions and our physical well-being. The science of emotions and health is strong, and the link is especially evident in the country's number-one killer: heart disease. The second connection is the pivotal one between the emotions and the other dimensions of health and longevity. It's quite possible that emotions serve as the main "connective tissue" tying together the nonbiological aspects of health with illness and longevity. That is, it could be that optimism, disclosure, social relationships, SEP, and finding meaning all affect health through their effects on emotions. Finally I will explore how the research on emotional longevity is connected to the future: Where is research in the new health science heading, what can we expect from it, and how should it be used?

Chapter 13

Emotional Connections

Some of my fondest adult memories of my mother and our time together are of our visits to her childhood home—New York City. My mother never missed an opportunity to return to her beloved New York. After my father passed, she would visit at least twice a year, always staying in the same hotel, where the staff knew her by name or sight. Walking through the lobby, she would be greeted with "Hello again" or "Welcome back" from bellmen and front-desk staff. After a little shopping her favorite pastime was to relax in her designated seat in the hotel lobby by a large window that looked out over bustling Sixth Avenue. There she could spend hours taking in the sight of the city and its inhabitants. The highlights of her visits were always the Sundays, when she would make her way up to Harlem to visit her "home" church, Bethel AME Zion, where not only did she discover her spiritual calling as a child and where she'd preached many times over the years, but also where she met my father.

In the summer of 1992, Elizabeth and I accompanied her on a visit to New York that we suspected might be her last. She was becoming frailer and weaker, and, in fact, on the day we were to leave for New York, she became acutely ill and couldn't travel. But she would not be denied this visit, and the next day she insisted on catching the first flight out. This trip was even more special than usual, since our extended family was throwing a major birthday party for her at the home of my cousin Elaine in Connecticut. Family members were coming not just from the New York and New England areas but from as far away as California and Colorado for the occasion. My mother was the spiritual matriarch of the family, and, knowing her health was failing, everyone wanted a chance to see her at least one more time. It was to be a really big event.

On the day of the party, as we were just starting our drive from Manhattan to Connecticut on the Major Deegan Expressway, we encountered one of those legendary New York traffic jams. Apparently there was a huge summer music festival in full swing in the area, with a carnival and parade that essentially shut down traffic on the expressway. Thirty minutes of not moving much at all turned into an hour of inching along, which turned into two hours of crawling forward. Sitting behind the wheel, I was having a major emotional meltdown. Here we were trying to get to what was in all likelihood my mother's last birthday party, with dozens of family members and friends waiting, and we were stuck in a traffic jam! I was on emotional overdrive—completely frustrated by the traffic, furious at myself for taking this route, and irritated that we would get to the party so incredibly late. I wanted this day to be absolutely perfect, and it was being ruined right before my eyes. I was probably more vocally angry than I'd ever been in my mother's presence. Whenever I would pound the steering wheel or bark my dismay, I would look at her in the backseat through my rearview mirror, expecting her to echo my frustration. After all, it was *her* birthday party we were missing. But instead what I got from her was not complicity with my mood, but just the opposite. In the mirror I saw in her the same serenity that I always saw. She was calm and relaxed, like nothing was happening. I can't remember exactly what she said during this seemingly (to me) interminable delay, but I do remember the main point: "Don't be concerned, everything will be fine, the traffic will clear. Let's just enjoy this time together."

Of course, I should have known that would be her response, but I was nevertheless amazed by her emotional control. That traffic jam *had* to register *something* negative in her at some point—my mother was not devoid of negative emotions like a Stepford robot. As children my brother and I had tested and confirmed her ability to show anger on various occasions. But for her the appearance of negative emotions was always short-lived, a cameo at best, and never the dominant sentiment. The traffic jam of course did end, and, as it turns out, many other family members had been caught in the same snarl coming out of New York, and many had arrived just ahead of us. As our car pulled up in front of Elaine's house, a throng of relatives, young children and elders, came to the car to greet my mother. Watching this wave of family affection flow around us had the same emotional impact as the closing scene of the movie *It's a Wonderful Life*, where all George Bailey's friends and family gather at his house with donations to save him from ruin. Similarly, the scene in Elaine's driveway was the quintessence of love. We were all enveloped by a powerful sense of deep affection, caring, and joy. A stark contrast to the state I'd been in on the drive there. And my mother was in her emotional element, an environment that reflected who she was.

Emotions. They hold a central, even exalted, place in the new definition of health. Like the other dimensions of health described in this book, emotions, along with stress, affect our physical well-being and longevity. In particular, the negative emotions of depression, anxiety, and anger have pervasive effects on everything from mortality and heart disease to the common cold. Because of this there's perhaps no better example than the negative emotions of the strong connections between the nonbiological and biological determinants of our health.

THE CONNECTIONS BETWEEN NEGATIVE EMOTIONS AND HEALTH

Any discussion of negative emotions has to start with what I call "the big three": sadness/depression, fear/anxiety, and anger/hostility. These emotions typically arise when an important goal is thwarted or is at least perceived as such. This includes goals related to maintaining personal safety and security, preserving valued relationships, or avoiding situations that are demeaning or offensive. Yet the phrase "negative emotions" does not imply that they are useless or should be avoided, since under the right conditions all emotions have their function. But when these emotions are experienced with great frequency or intensity, they can have profound consequences. When, for example, occasional sadness turns into depression, or when periodic apprehension turns into unrelenting anxiety, or when intermittent frustration turns into chronic anger and hostility. Let's briefly examine the nature of these negative emotions.

Sadness/Depression. In his book *Passion and Reason,* noted psychologist and author Dr. Richard Lazarus writes that sadness and depression are part of a family of emotions produced by unfavorable life circumstances, typically associated with loss or the perceived threat of loss.[1] In sadness the primary theme is irrevocable loss; that is, when sad, you feel that the loss cannot be reversed, which leads to a sense of helplessness. Depression, on the other hand, involves extreme sadness but is much more. One distinguishing feature of depression is the sense of hopelessness, the perception that the loss has pervasive negative implications for your whole life. Hopelessness can make life seem not worth living and, in extreme cases, can increase risk for suicide. The death of a loved one, the loss of material possessions, and even the loss of life goals or self-esteem are the kinds of losses that evoke sadness and depression. Major illnesses can also lead to extreme sadness and depression, since they can represent a loss of health or physical functioning.

Fear/Anxiety. Threats—such as to security, safety, survival, or identity—are considered the central components of fear and anxiety. Fear is the usual response to specific and sudden threats perceived as posing an imminent risk of death or injury. When confronted with this kind of threat, such as an out-of-control car jumping the curb and heading straight for you, specific regions of the brain are stimulated to allow quick action without thought.[2]

Anxiety, on the other hand, is the result of nagging, uncertain threats. Lazarus refers to anxiety as an existential emotion because of the vague nature of the threat and the uncertainty of whether it will occur or what we can do about it if it does materialize.[3] Anxiety manifests in a variety of ways, including as apprehension, nervousness, and worry. Anxiety disorders are extreme cases of debilitating fear and anxiety, including phobias, generalized anxiety disorder, posttraumatic stress disorder, and panic disorder.[4]

Anger/Hostility. Anger is typically the result of a demeaning offense that is a direct threat or that threatens someone or something of value. Aristotle defined anger as a belief that we, or our friends, have been unfairly slighted, which causes us both painful feelings and a desire or impulse for revenge. People also become angry when their impression of themselves as worthy, competent, and deserving is challenged. The resultant frustration leads to any number of emotions along the continuum of anger, ranging from irritation and annoyance to indignation and outrage to intense rage, fury, wrath, and hatred. Hostility is very much related to anger but is by no means the same. Hostility is not an emotion but rather is a personality, attitudinal, or cognitive (thinking) style, characterized by a tendency to view the world in ways that increase the likelihood that anger will be provoked. Hostile people are prone to see situations as involving personal offenses, and therefore they experience the emotion of anger frequently.

The number of studies exploring the effects of negative emotions on health is literally in the hundreds. Not all, naturally, are of sufficient scientific quality to draw firm conclusions. However, there have been a significant number of well-designed and rigorously conducted studies concluding that negative emotions participate in the development, course, and recovery phases of disease. The findings are especially strong regarding heart disease. This is fitting, given that heart disease is the number-one cause of death in industrialized countries. Researchers studying the links between negative emotions and heart disease have looked at a variety of cardiovascular outcomes. These include myocardial infarction (heart attacks), ischemic heart disease, congestive heart failure, and heart-disease-related death. Here is a synopsis of what we have learned about how negative emotions affect heart disease:

Depression predicts heart disease. Studies exploring the connection between depression and heart disease have examined depression in a number of ways. Many studies have looked at the relationship of diagnosed clinical depression to heart disease, but others have measured depressive symptoms such as sadness, hopelessness, feeling blue or downhearted, or loss of interest in pleasurable activities. Looking at symptoms of depression is important because many more people have such symptoms than have full-blown clinical depression. So it is critical that we know whether the latter group is also at risk for health problems.

There have been more than twenty long-term studies of depression and heart disease, involving over 25,000 participants. In these studies the level of depression is assessed at one point in time and participants are followed for anywhere from one year to as many as forty years.

When looked at as a whole, both clinical depression and a high number of depressive symptoms are associated with increased risk for heart disease. *Depression can lead to anywhere from a two- to fourfold increase in risk for heart disease.* Depression appears to be especially dangerous for people whose hearts have already been damaged from heart ailments. Depression in these individuals can cause dramatic increases in risk for fatal complications. But some studies have also found that depression can quadruple heart-disease risk over time in people who are initially healthy.[5]

Anxiety predicts heart disease. As with depression, studies of anxiety have measured anxiety in various ways. Some examined symptoms of anxiety such as worry, irritability, difficulty concentrating, or muscle tension. Others examined anxiety disorders such as generalized anxiety disorder, phobias, panic disorder, or posttraumatic stress disorder.

There have been more than ten long-term studies of the impact of anxiety on heart disease, involving more than 40,000 participants. These studies have demonstrated conclusively that either the presence of an anxiety disorder or a high number of anxiety symptoms is predictive of heart disease. Remarkably, *people with high levels of anxiety can have between two to seven times the risk of heart disease* compared to people with lower levels of anxiety. Anxiety is especially predictive of fatal heart disease.[6]

Anger and hostility predict heart disease. Three aspects of anger have been studied most rigorously in relation to heart disease.

One is the frequency with which angry feelings are evoked—called the experience of anger. The second is the expression of anger. Is the anger verbally or nonverbally expressed in some way, or is it inhibited or suppressed? The third is related to the concept of hostility. Hostility is more of a personality and attitudinal style by which a person views the world and other people in ways that increase the likelihood of anger. People who are high in hostility experience more anger than do others.

Findings on whether the experience of anger predicts heart disease are limited. Two long-term studies, lasting three and seven years, have found that people who report high levels of angry feelings are at exaggerated risk from heart disease. Among participants who were initially healthy, anger doubled their chances of having an initial cardiac event. Among patients with existing heart disease, *anger led to a sevenfold increase in their risk for a second cardiac event.*[7]

The way that anger is expressed may also affect the heart. At least two long-term studies have discovered that a tendency to hold in anger or a tendency to aggressively express it when provoked is predictive of cardiovascular disease.[8] Thus it appears that either extreme of anger articulation—extreme suppression or extreme expression—is bad for the heart.

Of all the components of anger, none has garnered more research attention in recent years than hostility. Hostility has many dimensions, among which is the greater tendency to become angry. But hostility is more than this. It is also associated with negative beliefs about and attitudes toward others. People who are hostile view others with cynicism, mistrust, and denigration. Hostile people expect others to be motivated by selfish concerns and that others are likely to be provoking or hurtful.[9] Dr. Timothy Smith of the University of Utah writes that hostile people exhibit "a devaluation of the worth and motives of others, an expectation that others are likely sources of wrongdoing, a relational view of being in opposition toward others, and a desire to inflict harm or see others harmed."[10] Over the last two decades, the health effects of hostility have been conclusively confirmed. Long-term studies indicate that hostility is predictive of both heart disease and other causes of death.[11] In one study researchers at Duke University found that hostility scores among medical students were predictive of the incidence of heart disease and all-cause mortality twenty-five

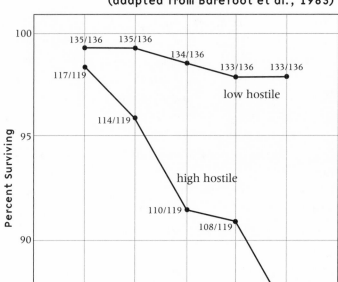

Figure 14: Survival as a Function
of Hostility Scores
(adapted from Barefoot et al., 1983)

years later. *Those participants with higher hostility scores had a nearly fivefold greater incidence of heart disease and, as depicted in Figure 14, over five times the death rate* of those with lower scores.[12]

The Connection Between Stress and Negative Emotions

It would be a major oversight to leave stress out of the mix when discussing the dimensions of emotional longevity. Stress is a prime elicitor of negative emotions; not the only one, but clearly one of the most powerful. Above and beyond its effects on emotion, stress is important because it is also a significant determinant of health and longevity. So what *is* stress? When I talk about stress, I'm not referring to a thing, I'm referring to a process. It's a process that begins with an interaction between you and your environment. An event occurs that might ultimately become a "stressor." *Whether* it becomes a stressor is determined by how you view it. The stress process involves four phases:

1. An event occurs in your environment, or you have thoughts about an event.
2. You appraise the event as threatening, taxing, or blocking some desired goal.
3. You appraise the event as exceeding your ability to cope with it, but you cannot avoid or alter what is happening, and you have difficulty reinterpreting the event in a less threatening way. It is then considered a stressor.
4. You exhibit a stress response that could include negative emotions, psychological or behavioral responses such as worry or withdrawal, and biological responses such as elevated blood pressure or impaired immune responses.[13]

As I was driving my mother to what would be her last birthday celebration, I had a classic stress reaction. The traffic jam was the environmental event that started the process. I appraised the event as blocking highly desired goals: to make it to the party on time and for the entire day to be perfect. The event exceeded my ability to cope with it—I could not avoid or alter the traffic and had difficulty accepting the fact that the day would *not* be perfect. My mother, on the other hand, viewed the situation quite differently. Even though the day was important to her, the traffic did not alter her emotional equilibrium, or at least not to the degree it did mine. Perhaps she relied on her characteristic optimistic explanatory style, knowing that traffic jams are never permanent, or maybe she leaned on her religious faith to find perspective and meaning, reciting in her mind the biblical scripture that "all things work together for good for them who love God."[14] Regardless of how she coped, the point is that events do not inevitably lead to stress reactions; it's our interpretation of events that does.

But the seemingly never-ending question has been, Can stress really cause illnesses? Not the occasional stress reaction when stuck in traffic but the reactions produced when stressors are extreme or unrelentingly chronic. Among scientists and laypeople alike, you will get an equal number of affirmative and negative responses to that question. Anecdotes about the stress/health connection abound. Elizabeth and I often talk about news reports of public figures who seem to succumb to an illness after enduring a long period of public reversal or a single tragic event. We often talk about the case of one of her admired sports figures, Coach Jim Valvano from North Carolina State University. Elizabeth is a rabid Atlantic Coast Conference (ACC) basketball fan, and the North Carolina State Wolfpack is one of her favorite teams. We lived in North Carolina during the devastating upheaval of Valvano's career, which garnered a lot of national press and made the first of many stark impressions of this kind on her. Valvano became a national celebrity when his underdog Wolfpack won the 1983 NCAA basketball tournament on a genuine buzzer-beater, pulling off one of the greatest upsets in college basketball history. Valvano's leaps of exhilaration around the court after winning became one of those classic television clips shown during college basketball season. He gave new meaning to "the thrill of victory."

A few years later Valvano was accused in a book of running a shoddy program plagued by drug use, improper booster support, point shaving, and poor grades.[15] The NCAA launched an investigation, and Valvano was raked over the coals of popular opinion for months. It was a sad, cruel event to witness. In the midst of it, he received a job offer from a school outside North Carolina but opted to remain in the state rather than uproot his children. Elizabeth was hoping he would take the job and leave the controversy behind. However, North Carolina State named him athletic director, too, and he remained. But in 1990 he finally resigned amid great controversy and pressure. He later became an analyst for ABC Sports, but his critics gave him little peace. None of the allegations were ever proved, but the team was put on probation for two years because the NCAA committee found that some players had sold shoes and tickets. Coach Valvano was not involved in any violations. Two years later, in 1992, he was diagnosed with bone cancer, and he died in 1993.

Elizabeth and I also talk about whether stress may have played a role in the illnesses of a Providence, Rhode Island, police officer, Major Cornel Young Sr. He lost his son, Sergeant Cornel Young Jr., himself a two-year veteran of the Providence police force, when he was shot by two fellow police officers as he tried to prevent a robbery while off duty. The case drew national media attention. Sergeant Young was a black officer who was killed by two white officers, and the incident has been attributed by many sources to racial profiling. Major Young, the highest-ranking black police officer in the Providence police force, was forced to walk a public tightwire of grief, job responsibility, and a need to determine if racism indeed played a role in his son's death. He remained on the force, was on inactive duty for a while, and even received a promotion. But Major Young was reportedly depressed, and friends were uncomfortable with his having a gun in his house. By the next summer he was diagnosed with Guillain-Barré syndrome and later with bone-marrow cancer. Major Young said that even the sympathy people offered became a problem, because he was besieged and felt he had no escape. When diagnosed with cancer, he asked publicly that he be granted privacy.[16]

Did stress contribute to the illnesses of coach Valvano and Major Young? We will of course never know. There is as yet little scientific evidence in humans that stress causes the onset of cancer. It very well may, but science has not provided clarity on this question. Yet there are formal case studies that strengthen the argument that stress can lead to other illnesses, some of which were noted by the legendary psychiatrist Dr. George Engel of the University of Rochester Medical School. Engel was perhaps first to coin the term "biopsychosocial" to describe how the many dimensions of health work together. In a very creative investigation, Engel once studied the factors surrounding sudden deaths using newspaper accounts over a span of six years. He selected 170 cases in which suicide could be ruled out as a possibility and the circumstances surrounding the deaths could be reconstructed. Engel discovered that a substantial number of the deaths could be linked to interpersonal losses, especially during the acute phase of bereavement. Here are some examples of what he found:[17]

- When told that her seventeen-year-old brother had died unexpectedly, a fourteen-year-old girl suddenly "dropped dead."
- Six months after receiving a clean bill of health following an electrocardiogram, a fifty-two-year-old man experienced a massive heart attack, which occurred the day after the funeral of his wife, who had died of cancer.
- An emotionally distraught eighty-eight-year-old man died of acute pulmonary edema shortly after learning of his daughter's death.
- An eighteen-year-old girl who had been reared by her grandfather died suddenly and without warning upon learning of his death.

You may know of similar stories from your own personal history or your local news. These are intriguing and provocative but fall under the category of anecdotal data. The circumstantial evidence in these stories is compelling, but it is always hard to say for sure what the exact cause of an illness or death is in a particular individual. It could be stress, or it could be something else. But science has finally moved beyond the anecdotal and begun to confirm the everyday notion that stress is tied to our health and longevity.

Scientists explore the link between stress and health in many different ways. Some examine what are called stressful life events: the frequency of occurrences that most of us would consider stressful, such as divorce, relocation, serious illness, the death of a loved one, or financial difficulties. Others focus on the specific types of potential stressors, such as caring for a loved one with Alzheimer's disease, enduring job stress, having a history of child abuse, experiencing bereavement, or even surviving earthquakes and war conditions. Still other researchers assess stress more subjectively, focusing more on individuals' appraisals of stressors in their lives. The link between stress and health has also been explored by studying the long-term health effects of stress-reduction approaches.

Stressful Life Events and Health

Stressful life events predict mortality. In Sweden researchers assessed the frequency with which ten different stressful life events occurred within a twelve-month period for 750 healthy men undergoing routine medical examinations. They recontacted the men after seven years to determine their health status. The men who had three or more events, which was considered a high-stress profile, were significantly more likely to have died during the seven years than were those reporting no life events. In fact, the *men scoring higher in stressful events were three times more likely to die* compared to those with lower stress scores. The differences in mortality between the two groups remained statistically significant even after taking into account smoking, social class, social support, and other factors.[18]

In a national study of 2,323 heart-attack survivors published in the *New England Journal of Medicine,* researchers examined the association between stressful life events and death during a three-year period. As shown in Figure 15,[19] *heart-attack survivors who reported*

Figure 15: Death Rates as a Function of Life Stress and Life Stress Plus Social Isolation
(adapted from Ruberman et al., 1994)

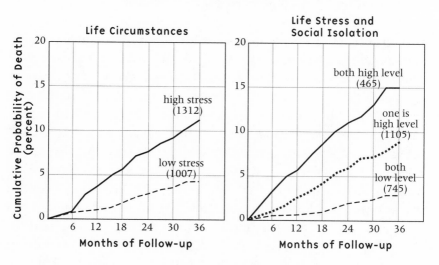

higher levels of life events were twice as likely to die during the three years than were those with lower levels of stressful events. *Patients who had higher life stress and were socially isolated from friends and family were four times more likely to die* than were those who were low in stress and social isolation. This study illustrates how some of the determinants of health can together have a synergistic effect. The effects of stress and social isolation combined were greater than those of either one alone.

Stressful life events predict disease progression. Stressful life events may also adversely affect the course of a disease already present or worsen the clinical features or symptoms of the illness. One of the best examples of this phenomenon is seen in the research on stress and AIDS. Dr. Jane Leserman and associates at the University of North Carolina at Chapel Hill have conducted several studies showing that, among people who are HIV positive, those who had experienced greater recent stress in their lives were significantly more likely to develop AIDS. In one study 82 HIV-infected patients without illness symptoms or AIDS were seen every six months for up to five and a half years. Leserman discovered that *stressful life events were related to a faster progression from HIV-positive status to the development of full-blown AIDS. People who had more life-events stressors were twice as likely to develop AIDS* as were those lower in such events.[20]

Stressful life events predict vulnerability to illness. Dr. Sheldon Cohen of Carnegie Mellon University in Pittsburgh has conducted a truly elegant series of studies that convincingly demonstrate that psychological stress increases our vulnerability to illness—in this case our susceptibility to the common cold. In one study, published in the *New England Journal of Medicine*,[21] Cohen's research team studied 394 healthy volunteers who completed questionnaires about stressful life events and their perceptions that current stressors exceeded their ability to cope. Participants were then exposed, via nasal drops, to one of five different respiratory viruses. The volunteers were quarantined and monitored for evidence of infection or cold symptoms. Remember, not everyone exposed to a cold virus gets infected or comes down with a cold. As depicted in Figures 16 and 17, Cohen found that both

Figure 16: Association Between Psychological Stress and the Rate of Colds
(adapted from Cohen et al., 1991)

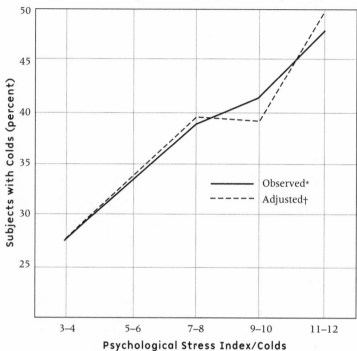

* Observed = percent of colds without accounting for control variables
† Adjusted = percent of colds after accounting for more than 12 control variables, including age, sex, education, allergic status, smoking, diet, and white blood cell counts

respiratory infections and colds increased in a dose-response manner with increases in stress.[22]

The Health Consequences of Work Stress, Bereavement, and Caregiving

Frequent stressful life events are not the only stressors that affect health. Exposure to a single stressor, if intense and prolonged, can also be toxic. Three stressors that fit this category are job strain, bereavement, and caregiving.

Figure 17: Association Between Psychological Stress and the Rate of Viral Infections
(adapted from Cohen et al., 1991)

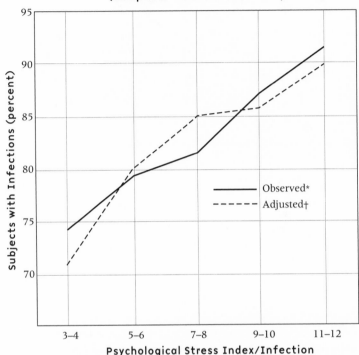

*Observed = percent of infections without accounting for control variables
†Adjusted = percent of infections after accounting for more than 12 control variables, including age, sex, education, allergic status, smoking, diet, and white blood cell counts

Job strain predicts heart disease. Everyone experiences some stress at work, but research has isolated key dimensions of the work environment that are especially stressful, resulting in what researchers call "job strain." Job strain consists of two components: the level of a job's demand and the degree of decision latitude it affords. Job demand is the obvious one. Work that is hectic and psychologically taxing is considered high in demand. Another, less obvious component of job strain is decision latitude, which is the degree to which the job provides decision making freedom and control over how the work is done. Decision latitude includes the degree of personal freedom associated with the work, such as an employee's ability to make a personal phone call and how much intellectual discretion is involved in getting the job done. Jobs that are more repetitive in nature usually do not allow much discretion or intellectual creativity. Jobs that require more formal education often allow more latitude. Research from around the world indicates that *high job strain can increase the risk of coronary heart disease two- to fourfold.*[23]

Bereavement predicts mortality. The loss of a loved one can be devastating. Bereavement produces unrivaled emotional pain, involving symptoms of shock, intense sadness, anger, yearning, despair, guilt, humiliation, disbelief, and confusion. As difficult as it is, most people manage to survive bereavement without long-term emotional or biological damage. Most of us are able to carry on with life, although the loved one is still missed and some degree of mourning may never fully disappear. For some, bereavement can be an acute stressor whose effects eventually subside. For others, however, the emotional toll of bereavement is thoroughly devastating and long-lasting. In these cases bereavement becomes a chronic stressor that elicits profound biological changes. This is the reason that bereavement is a risk factor for morbidity and mortality, especially in the weeks and months immediately following the onset of widowhood.

Men and, interestingly, younger widows and widowers are especially susceptible to grief-related mortality following the death of a spouse. In a study of over 1 million married people, ages thirty-five to eighty-four in Finland, men and women who had lost spouses were studied for five years. Bereavement was shown to have more impact on bereaved men than on the women. Bereaved women had a 6 percent higher death rate than nonwidowed women of the same age. Bereaved men, however, had a 17 percent higher death rate than their nonwidowed counterparts.[24] Is stress the cause of the excess mortality in the widowed? It may not be the only reason, but it certainly figures prominently.[25] Another factor is that the surviving spouse may lose economic resources and the instrumental support associated with everyday tasks such as cleaning, food preparation, child care, and taking medication. However, these losses are unlikely to be the cause of the subsequent death during bereavement of the surviving spouse. The effects of losing economic support and task-related assistance would most likely have a cumulative effect, worsening over time. Deaths during bereavement are most likely to occur in the short term rather than the long term.

More likely explanations involve the emotional stress and loss of emotional support during bereavement. The high rate of survivor death within six months after spousal death suggests a causal role for the intense emotional fallout of bereavement. The fact that younger widows and widowers are at greater risk than those who are older also implicates stress as a causal factor. For older couples the death of a spouse is considered an "on-time" event. It is something that is expected to occur at some point among the elderly. For younger people the death of a spouse is "off time," something that is not supposed to happen and is therefore perhaps more shocking and harder to cope with. Finally, when scientists have looked at the exact medical causes of death during bereavement, several of those causes are sensitive to stress. These include heart disease, alcohol-related diseases, suicide, motor-vehicle accidents, violence, and lung cancer (tied to cigarette smoking).

Caregiving for a disabled relative predicts illness and death.
A great deal of stress research in recent years has focused on a large and rapidly growing segment of society—people who care at home for a relative with a disability or chronic illness. Most caregivers are spouses or adult children of people with illnesses such as Alzheimer's disease, stroke, heart disease, and other problems that impede their ability to completely care for themselves. It has been estimated that as many as 15 million people serve as caregivers, many of whom are elderly themselves.[26] Although caregiving may be emotionally rewarding, caregivers on the whole report extraordinarily high levels of stress and burden.[27] A recent long-term study of caregivers looked at caregiving's effects on their mortality. The study, published in the *Journal of the American Medical Association,* followed 392 elderly people who were providing care for a spouse. They were compared with a group

of more than 400 noncaregivers of similar age. Over a four-year period, the ***caregivers who were experiencing mental and emotional strain were 63 percent more likely to die*** during the four-year study than were the noncaregivers. Caregivers not reporting strain were not at increased risk of death.[28]

Chapter 14

Emotions as "Connective Tissue"

The basic premise of this book is that health is more than the absence of disease, that our biological status is only one of several broad categories of factors that influence our health and longevity. Biology operates in conjunction with thoughts and actions, with social and economic factors, with existential/religious/spiritual factors, and with the emotional factors just described. Like the other dimensions, emotions are strongly linked to health outcomes. But emotions hold a special place in this new definition of health. The very title of this book implies that emotions hold an exalted position with respect to longevity—kind of a superior among equals. I take this position not because emotions are somehow more powerful determinants of health than the others. The main reason for the elevated status of emotions is this: Emotions serve as a key pathway by which the other factors affect health. All the other examples of the dimensions of health discussed in this book—optimism, disclosure, social relationships, economics, finding meaning—may well have their effects on health *because they first have an impact on emotions.* Emotions are a crucial part of the "connective tissue" that binds the other determinants to longevity.

This is not to suggest that emotion is the *only* pathway or connective tissue linking the other dimensions with longevity. Indeed, there is strong scientific justification for the idea that many of the dimensions described herein also affect other potential pathways to longevity. For example, optimism or social support affect activities such as abstaining from smoking or excessive alcohol intake, getting regular exercise, eating a healthy diet, getting sufficient sleep, having regular physical examinations, and even taking medication as prescribed. In addition, emotions cannot by themselves *completely* account for why things like SEP or social relationships or religious participation are so health enhancing. Emotions, even when they act as a key pathway to longevity, rarely act alone. People who are high in negative emotions also exhibit other factors that might put them at risk for illness. But the emotions are unique in that they are the *one pathway that all the dimensions of health share*. Emotions, especially negative emotions, are affected by each of the other five dimensions of health. Emotions are the common threads that run through all the stories of people profiled in this book, who benefited from close social relationships, or were able to find meaning, or were able to disclose traumatic experiences. The one outcome that all these people shared was a greater sense of emotional well-being. Every dimension might not affect smoking or visiting the doctor, but every dimension does in fact alter our emotions. Emotions, like biology, might be considered "superdimensions" of health, since emotions are a shared route down which the others travel to affect health.

There is another reason for the special place reserved for the emotions, which again demonstrates the connections between emotions and the other dimensions of health. The other dimensions can be viewed as protectors of our emotional well-being. They do this not necessarily by preventing adversity from occurring but by buffering us from its untoward emotional effects. In fact, it is during times of adversity that the connection between the dimensions of health and the emotions is seen most vividly. When adversity comes into our lives, those of us who, for example, are more optimistic, better able to disclose feelings, or able to find meaning have a greater degree of emotional protection. So, as depicted in Figure 18, people who are able to better utilize these characteristics have a buffer zone between their emotions and adversity.

There is one more reason that emotions are so pivotal to our overall health. Emotions act as a kind of well-being thermometer, providing us with a reading of how effective our buffers have been in protecting us from stress or if they themselves are the source of some difficulties. That is, our emotions at any given time reflect, among other things, the sum total of how optimistic we've been, the conse-

Figure 18: Dimensions of Health as a Buffer Zone Between Adversity and Negative Emotions

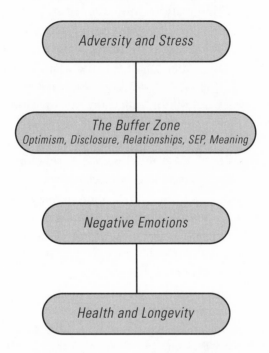

quences of undisclosed traumas, whether there are problems in our social relationships, whether we're experiencing economic difficulties, whether we've been able to find meaning in some unavoidable negative life experience. Like a thermometer displaying temperatures, our emotions give us a reading about the state of the other dimensions of health and the state of our lives more generally: Are circumstances dangerous or safe, threatening or wondrous? Have good things come into our lives, or have we lost something or someone valuable? When our emotions become exceedingly negative, they are signaling to us that there are problems in some aspect of our lives, problems that could likely be addressed in part by greater attention to the dimensions of our health in the buffer zone.

The proof of the connections between emotions and the other dimensions of health has already been described in each of the previous chapters. But by way of summary, below is what has been discovered.

Looking on the bright side: optimism connects with emotions. In Chapter 1, I talked about optimism being defined as a person's expectations about the future (dispositional optimism) and explanations of the past (explanatory style). Both types of optimism are known to defuse negative emotions and heighten positive emotions. This connection has been observed in research with participants as varied as naval recruits, law and undergraduate students, bypass-surgery patients, and people from many other populations. In addition, perhaps the clearest demonstration of the connection between the explanatory-style form of optimism and emotions is the research on depression. Persons with diagnosed clinical depression tend to use a pessimistic explanatory style for bad events, and the more severe the depression, the more pessimistic the explanatory style. The depression lessens when therapy produces changes in the explanatory style from pessimistic to more optimistic.[1] Finally, in addition to its direct effects on mood, optimism has been demonstrated to be a powerful buffer between mood and adverse life experiences, such as infertility, the stress of providing care for an ill relative, cancer, bone-marrow transplantation, and AIDS.

Revealing and dealing: disclosure connects with emotions. People who are able to disclose traumatic experiences have improved long-term health compared with those who conceal such traumas. Interestingly, revealing traumatic experiences often leads to a short-term worsening of mood, since the act of disclosing long-repressed negative events can be emotionally jarring. But over time disclosing the trauma produces an improvement in mood and in health. Although not every study of disclosure has produced these results, a comprehensive examination of this research indicates that writing about traumatic events does indeed lead to a reduction in emotional distress.

Getting by with a little help from friends: social support connects with emotions. People who have larger social networks and supportive friends and family not only live longer on average but experience greater emotional well-being. In particular, the research demonstrates a strong link between higher levels of social support and lower levels of depression. What is really striking about supportive social relationships is their ability to serve as a buffer that protects individuals from the emotional consequences of stress.[2] A study in Sweden discovered that life stress was a significant predictor of mortality from heart disease, cancer, and alcohol-related health problems in middle-aged men. In that same study it was also discovered that the presence of social support blocked the deadly effects of stress. For men with high emotional support, stress was not associated with death— that is, emotional support had a buffering or protective effect against health consequences associated with stress. For men low in emotional support, stress was especially deadly. For participants who had more than two major stressful life events, the risk for death was fifteen times higher than it was for men with no stressful life events. Stated differently, among men with low emotional support, those with zero life stressors at baseline had a death rate of only 1.5 percent, but those with more than two stressful life events had a death rate of 25.9 percent.[3]

Climbing a little higher: socioeconomic position connects with emotions. The SEP/health gradient is as much a reality for emotional health as it is for physical health. People who are lower in SEP experience higher levels of depression, anxiety, and anger-provoking situations than do their higher SEP counterparts.[4] At this time it is hard to say exactly why this is the case. It could be that people of lower SEP experience more untoward life events or, alternatively, that higher SEP acts as a buffer, protecting individuals from stress. It is probably a little bit of both. As noted by Dr. Andrew Baum of the University of Pittsburgh, lower SEP "is likely to be correlated with settings with higher population density, noise, crime, pollution, discrimination, poor access to resources, and hazards or privations. Limited income, education, and/or lower social class may cause people to live in poorer, stressful settings or may perpetuate their living in such areas. Higher [SEP] communities appear to have fewer hazards or privations, more support, and be able to afford more options for coping with problems that do occur."[5]

A journey of discovery: finding meaning connects with emotions. In the chapter on finding meaning, I reviewed a number of studies that connected this process with emotions. These studies clearly illustrated the buffering effects of finding meaning. That is, people who were able to find meaning following an aversive life experience reported lower levels of depression and stress than did others. Recall, for example, that among men who were combat veterans of World War II and the Korean War, those who were able to find the most positive effects from serving in the military and experiencing combat were the least likely to experience combat-related stress symptoms. The more the men viewed their military experience as undesirable, the more they experienced stress symptoms. These findings remained significant after accounting statistically for the degree of combat exposure. In a study of recently bereaved individuals, those who were able to find meaning showed better adjustment in the months after the death than did those who were less able to find any meaning in their loss. Finally, among mothers of preterm infants who had been in neonatal intensive care, those mothers who became more emotionally distressed over the ensuing eighteen months after discharge tended to be those who failed to seek meaning in or find benefits from the crisis.

Final pathways: biology connects with emotions. As in other parts of this book, the question always arises as to how it is that a non-biological dimension of health leads to disease and death. And like the other dimensions described earlier, negative emotions and stress likely affect health and longevity through their effects on biology and, to a certain degree, behavior. These connections are most clearly illustrated in research on stress. Exposure to stress has been demonstrated to alter every system of the body, including the immune, cardiovascular, neuroendocrine, and central nervous systems.[6] In the short run, when stress is acute, such biological changes may actually be adaptive, helping us adjust to or effectively address the situation at hand. Over the long run, however, when stress becomes more chronic, these same bodily changes can occasion the onset of disease, exacerbate existing illnesses, or slow our recovery from an illness. Stress may also lessen health-promoting behaviors such as exercise and sleep and increase health-damaging behaviors such as smoking, unhealthy dietary practices, and excessive alcohol intake.[7]

A good example of the deadly biological changes associated with stress is the research on myocardial ischemia, which is a common feature in many patients with coronary heart disease. Myocardial ischemia is characterized by a narrowing of the coronary arteries and a reduction of blood flow to the heart, which places a person at risk for heart attacks or other deadly cardiac events. It has long been recognized that physical stress such as exercise can increase ischemia, which is one reason cardiologists often use a treadmill test to assess it. Recent studies now indicate that psychological stress—produced by performing mental arithmetic or making an impromptu speech—can produce a similar onset of ischemia. Patients who have a history of heart disease and who show the most ischemia in response to psychological stress are at greatest risk for a second heart attack or other cardiac abnormalities.[8] The good news is that among patients who exhibit such ischemia, stress-management techniques can be used to prevent subsequent heart attacks.[9]

While emotions may be the connective tissue linking nonbiological factors with health, there are many other connections among these dimensions. That is, optimism, social relationships, socioeconomic position, and the process of disclosing trauma or finding meaning are also connected to each other. Thesese connections can lead to a deadly cascade of experiences. For example:

- People who are lower in SEP have lower levels of social support.
- People who are lower in social support are less likely to disclose trauma.
- People who are less likely to disclosure trauma are probably less likely to find meaning in it.
- People who are less likely to find meaning are probably less likely to be optimistic.
- People who are less optimistic are more likely to experience negative emotions.
- People who experience more negative emotions show increased risk for illness and death compared to those who experience fewer such emotions.

The causal chain, of course, can work in many directions and have different entry points and routes. Social dimensions of health affect psychological and emotional dimensions, but the reverse is also true. Our emotional state can influence our relationships with others. The interactions and connections among the dimensions of health are limitless. For example, although pessimism often causes negative emotions, those emotions may in turn reinforce pessimistic beliefs. Those pessimistic beliefs can alienate others, leading to diminished social support and fewer opportunities to disclose stressful life experiences.

Although much of the research I have described examines one element or causal factor at a time to determine its health effects, in reality all the dimensions are connected to each other in very dynamic ways. In fact, risks for illness and death increase or decrease depending on the accumulation of risk or protective factors across the dimensions. For example, risk for death is greater for people who are depressed *and* have a low income *and* are socially isolated than for people with only one of these characteristics.[10] Another example of these cumulative effects is the study I described in the previous chapter, where heart-attack patients reporting higher levels of stressful life events were twice as likely to die over three years than were those with lower stress levels. However, those who were high in stress *and* were socially isolated were *four times as likely to die* than were those lower in stress and isolation.[11] You might also recall from the introductory chapter that ulcers are most likely caused by a *combination* of stress and a bacterium.

Chapter 15

Connecting to the Future

What does the future hold for new health science approaches to research on emotional longevity? What progress can we expect over the decade? The short answer is accelerated progress within each of the nonbiological dimensions of health, with further understanding of how they relate to each other and to biological processes. The investment and commitment to this field by the NIH and Congress is growing, and scientists from different disciplines are finding more common ground every day—even between social scientists who study communities and biological scientists who study genes. But what are the newest, most groundbreaking directions this research could take, and what are the most provocative long-term implications from it? And how will the new discoveries affect our lives?

When it comes to the future of emotional longevity, I have both predictions and hopes. Below are a number of predictions about what will be the most influential discoveries from the new health science over the next ten years. These are not outrageous pie in the sky predictions. They're based on research that is already under way and already bearing fruit.

POSITIVE EMOTIONS AND RESILIENCY WILL BE DISCOVERED TO AFFECT LONGEVITY

The impact of negative emotions on health has been more systematically investigated than postitive emotions, but this does not mean that positive emotions are unimportant. There is now great scientific energy focused on discovering the health effects of positive emotions and attitudes, and psychological processes such as joy, love, gratitude, interest, contentment, amusement, and forgiveness. Scientists have already discovered that positive emotions and attitudes can counter and undo some of the physiological effects of negative emotions. Positive emotions help people think more flexibly, creatively, and openly in problem-solving situations.[1] I predict that positive emotions will be powerful predictors of lower rates of illness and death. People who experience more frequent positive emotions will have lower death rates than others. I further predict that research will discover psychological and behavioral techniques for fostering positive emotions for those who experience them less, and the learning of such approaches will prevent or speed the recovery from disease. The work on positive emotions and attitudes will foreshadow an even larger scientific trend toward the study of "resiliency." By resiliency I am referring to the processes that lead to good outcomes in the presence of threats to one's overall well-being.[2] Most of the chapter topics in this book are examples of factors that foster resiliency, such as optimism, social support, disclosure, and finding meaning. I predict that there will be an acceleration of discoveries about how we can become more resilient in the face of unfortunate life experiences. This work is already beginning to explore, for example, why some people have coped better than others following the September 11th attacks.[3]

NONBIOLOGICAL FACTORS WILL BE DEMONSTRATED TO AFFECT GENE EXPRESSION

Perhaps the ultimate demonstration of connections will be research that ties nonbiological factors with gene expression; that is, the turning on or off of genes. Determining which factors truly contribute to gene expression has obvious and important implications for the treatment and prevention of disease. How could factors such as emotions, stress, expectations, and social relationships affect the activation of genes inside the nuclei of cells? The logic for this is fairly straightforward. As described throughout this book, these nonbiological factors affect hormones like cortisol, epinephrine, norepinephrine, and many others. These hormones in turn activate other substances inside cells, which in their turn activate the expression of genes inside cell nuclei. There is already strong indication from basic research that environmental and behavioral factors (e.g., stress, exercise, social interactions) affect gene expression through the activation of hormones,[4] and clinical research on this topic is emerging. For example, researchers are now discovering how stress may adversely affect the expression of genes associated with the immune system.[5] I predict that scientific collaborations between behavioral and social scientists and geneticists will conclusively demonstrate the close connections between all of the dimensions of health with gene expression. These dimensions will be shown to affect the expression of genes associated with heart disease, diabetes, cancer, arthritis, and other chronic diseases, and even with the structure and functioning of our brains.[54]

NEW NONDRUG INTERVENTIONS WILL BE DISCOVERED THAT CAN PREVENT AND TREAT DISEASE AND OTHER ADVERSE CONDITIONS

Most of the research I have described has been of the causal nature—that is, research showing the links between the dimensions of health and disease outcomes. We now know that emotional, psychological, and social factors contribute to disease. But one of the next frontiers for the new health science is the discovery of a wider array of interventions that, based on the findings from research, can actually prevent or treat disease. This work has already begun and has led to many nondrug interventions to prevent disease and treat chronic illnesses and adverse conditions, including arthritis, asthma, heart disease, diabetes, cancer, child abuse, low birth weight, and other problems.[6] I predict that there will be a greater preponderance of randomized clinical trials demonstrating that behavioral, psychological, emotional, or social interventions can prevent the onset of a variety of illness and conditions. Scientists will also discover that many of these nondrug interventions will be as effective as drug treatments and that in many cases the combination of drug and nondrug treatments is optimal.

THE NEW NONDRUG INTERVENTIONS WILL ALSO BE SHOWN TO DRAMATICALLY CUT HEALTH-CARE EXPENDITURES

The new nondrug approaches to treat or prevent disease will be shown to produce significant declines in health care expenditures. Scientists will incorporate more exacting cost analyses into their studies of preventive and treatment interventions. These analyses will conclusively demonstrate that by paying for interventions that enhance the emotional or social well-being of the people they serve, health-insurance companies actually *save* money. Such savings will result from decreased health-care utilization and less need for higher cost medical and surgical interventions.[7]

THE SOCIAL ELEMENT OF HEALTH WILL BE DISCOVERED TO BE MORE IMPORTANT THAN PREVIOUSLY IMAGINED

Although every element of health will show accelerated growth and progress over the next ten years, I predict that research on the social element will be especially noteworthy. Two trends will be responsible for this. The first is the greater emphasis on racial and ethnic-based health disparities. Wide differences in health outcomes and life expectancy exist among the major racial and ethnic groups in this country.[8] In response, the NIH and other components of the Department of Health and Human Services have made understanding and eliminating health disparities a top priority. Every institute of the NIH has developed a strategic plan to increase its research funding in this area.[9] The social element of health, along with the cultural factors subsumed therein, are critical to understanding racial and ethnic-related disparities. The second trend has to do with the fact that health-related social-science research is rapidly moving beyond the study of interpersonal processes to an exploration of the health impact of larger social systems, government policies, economic systems, media influences, neighborhoods, and institutions such as schools, the health-care system, religious organizations, and the workplace. Individuals are both affected by and play a role in shaping these larger social systems. The health, behavioral, and biological consequences of the dynamic interaction between individuals and social systems will be a source of exciting discoveries in this next decade.[10]

A VISION FOR THE FUTURE

Based on the findings described in this book and trends like those above, it is my hope that we will have both the means and the will to expand our approach to health—on both an individual and a societal level. Such a change will mean that:

- As individuals we will routinely consider the emotional, psychological, and social dimensions of our health, in addition to the biological and behavioral, in our efforts to prevent or recover from illness.
- People at genetic risk for illness will utilize behavioral, psychological, and emotional, or social techniques to prevent the activation of disease-contributing genes.
- Medical schools will update their curricula to include more about the new health science so that physicians in training will learn as much about the nonbiological dimensions of health as about the biological.
- Health assessments in medicine will include the nonbiological evaluations alongside the biological so that treatments may be targeted at the correct dimension(s) of the problem.
- Health-insurance plans will pay for treatments aimed at nonbiological dimensions of illness for interventions by health professionals (including psychologists, nurses, and others) to keep people healthy or to help them recover more rapidly from illness.
- As a society we will have a greater recognition of the manifold influences of the social context, including how families, neighborhoods, schools, work sites, and economics affect health and lifestyle factors such as exercise patterns, diet, and smoking.
- Public health, health promotion, and disease prevention will become greater priorities in society. Health-promotion and disease prevention will become routine considerations among those who set policies in the workplace, in school settings, and in health care, and they will become increasing priorities in federal research funding.

Finally it is my hope that these predictions and hopes will make available to everyone what my mother seemed to have had naturally—harmony of mind, body, and spirit.

Notes

INTRODUCTION: EMOTIONAL LONGEVITY: TOWARD A NEW DEFINITION OF HEALTH

1. Rowe and Kahn (1998).
2. The term "nonbiological" is not used to evoke a mind/body dualism but as a convenient way to describe aspects of human functions that are not synonymous with physiology. However, it is recognized that all aspects of human functioning have a biological component. I will make clear throughout this book that the nonbiological dimensions are inextricably linked to the biological.
3. Anderson, N.B. (1998).
4. Lichtenstein et al. (2000).
5. Surwit et al. (1984).
6. Grant, Piotrowski, Chappell (1995); Kaplan and Lusky (1988); Kaplan and Camacho (1983); Idler and Angel (1990); McGee et al. (1999); Schwartz et al. (1999).
7. Schwartz et al. (1997); Low-Beer et al. (2000); Wagner et al. (1995).
8. Preamble to the constitution of the World Health Organization as adopted by the International Health Conference, New York, June 19–22, 1946; signed July 22, 1946 by the representatives of sixty-one states (official records of the World Health Organization, no. 2, p 100) and entered into force on April 7, 1948. The WHO definition also stated that health was not merely the absence of disease.
9. The expression "expanded view of health" has been used by many scientists and organizations, most notably the Center for the Advancement of Health in Washington, D.C.
10. Larson (1996).

11. Many scientists and scholars have written about this perspective over the years, using terms such as "behavioral medicine," "psychomatic medicine," "health psychology," or the "biopyschosocial," "biobehavioral," or "biosocial perspectives."
12. The terms "interdisciplinary" and "transdisciplinary" are often used interchangeably.

PART I: THOUGHTS AND ACTIONS 1: EXPECTATIONS, EXPLANATIONS, AND BELIEFS

1. Interview with Art Berg (2001). Berg is president of Invictus Communications and a member of the National Speakers Association, where he has earned the highest designation of Certified Speaking Professional, an honor achieved by less than 7 percent of its members. He was awarded a SuperBowl ring by the Baltimore Ravens for his role in motivating them to their championship. Books by Berg: *The Impossible Just Takes a Little Longer* and *Finding Peace in Troubled Waters.*

CHAPTER I: EXPECTATIONS AND EXPLANATIONS

1. Scheier, Carver, Bridges (1994); Marshall et al. (1992).
2. Greenberg and Springen (2001).
3. Carver and Gaines (1987).
4. Bromberger and Matthews (1996).
5. Scheier et al. (1989).
6. Scheier et al. (1999).
7. Schulz et al. (1996).
8. Segerstrom et al. (1998).
9. Raeikkoenen et al. (1999).
10. Personal communication from Dr. Martin Seligman (2001).
11. Seligman (1991).
12. Zullow et al. (1988).
13. Zullow and Seligman (1990).
14. Peterson and Seligman (1984).
15. Seligman et al. (1988).
16. Nolen-Hoeksema, Girgus, Seligman (1992).
17. Peterson, Seligman, Vaillant (1988).
18 Kubzansky et al. (2001).
19. Peterson et al. (1998).

20. Maruta et al. (2000).
21. Kamen-Siegel et al. (1991).
22. Segerstrom et al. (1996).
23. Kubzansky et al. (in press).

CHAPTER 2: IS OPTIMISM ALWAYS GOOD?
IS PESSIMISM ALWAYS BAD?

1. Personal communication from Dr. Christopher Peterson.
2. Seligman (1991).
3. Snyder and Dinoff (1999).
4. Sheier et al. (1989); Fitzgerald (1993); Carver et al. (1993); Fry (1995); Stanton and Snider (1993); Struffon and Lumpkin (1992); Fontaine, Manstead, Wagner (1993); Litt et al. (1992); Taylor et al. (1992).
5. Carver and Scheier (1999).
6. Charles Gibson interviewed Mattie Stepanek, his mother, Jeni Stepanek, and President James Carter on *Good Morning America*, December 4, 2001; Chris Cuomo reporting. President Carter is an idol of Mattie Stepanek's, but they had never met although Carter had written the foreword to one of Mattie's books of poems. Carter surprised him with a visit during the interview.
7. ABC News Internet Ventures, "A Purpose of Peace. Best-Selling Boy Poet Promotes Peace," November 16, 2001.
8. Lyman (2001).
9. Norem and Cantor (1986a).
10. Ibid.
11. Norem and Cantor (1986b).
12. Ibid.

CHAPTER 3: THE POWER OF BELIEFS
AND ILLUSIONS

1. Seligman (1995).
2. Interview with Dr. Albert Bandura by Jill Kester in *The Observer*, a newsletter of the American Psychological Society (2001).
3. Bandura (1997).
4. Ibid.
5. Interview with Mr. Wally Amos (2001). Self-help and inspirational books by Wally Amos include *The Power in You, Man with No Name, Water-*

melon Magic, and *The Cookie Never Crumbles.* For more information, visit www.wallyamos.com and www.unclewally.com.

6. Bandura (1997).
7. Bandura et al. (1999).
8. O'Leary (1985); Clark and Dodge (1999); Bandura (1997).
9. Cacioppo (1994); Cacioppo (2000); Sapolsky (1998); Lovallo (1997).
10. Gerin et al. (1995); Bandura et al. (1985); Wiedenfeld et al. (1990).
11. Seeman et al. (1996).
12. O'Leary et al. (1988); Smarr et al. (1997); Lorig and Holman (1993).
13. Bandura, Reese, Adams (1982); Bandura, Adams, Beyer (1977).
14. Sanderson, Rapee, Barlow (1989); Litt, Nye, Shafer (1995); Litt, Nye, Shafer (1993).
15. Manning and Wright (1983); Bandura et al. (1987); Dolce (1987); Clark and Dodge (1999).
16. Bandura (1997).
17. *National Sports Report,* Fox News; Kevin Frazier reporting, November 2001.
18. Taylor and Brown (1994); Taylor and Brown (1988).
19. Alloy and Abramson (1979).
20. Lewinsohn et al. (1980); Coyne and Gotlieb (1983); Ruehlman, West, Pasahow (1985).
21. Taylor and Brown (1988).
22. Taylor (1989).
23. Taylor (1983).
24. Taylor et al. (2000).
25. Taylor et al. (1992).
26. Kübler-Ross, E. (1969, 1987).
27. Reed et al. (1994).
28. Kübler-Ross (1987).
29. Reed et al. (1999).
30. *National Sports Report.*
31. Kabat-Zinn et al. (1998); Kabat-Zinn et al. (1985); Miller, Fletcher, Kabat-Zinn (1995); Kabat-Zinn (1995).

PART II: THOUGHTS AND ACTIONS 2: CONCEALING AND REVEALING TRAUMA

1. Smyth (1999).
2. The volunteer sample included 61 people with asthma and 51 with arthritis; 107 completed the study—58 in the asthma group and 49 in the rheumatoid arthritis group.

3. Lung function was measured by the mean percentage of predicted forced expiratory volume (FEV) in one second.
4. Matarazzo (1984).

CHAPTER 4: SILENCE, SECRETS, AND LIES:
THE HIGH PRICE OF CONCEALMENT AND AVOIDANCE

1. Interview with Linda Ellerbee (2001).
2. Horowitz (1975); Lepore (1997); Lepore (on press).
3. Wegner (1997).
4. Wegner et al. (1997); Wegner and Eber (1992); Wegner (1997).
5. Wegner, Eber, Zanakos (1993); Beevers et al. (1999); Wegner and Zanakos (1994).
6. Ibid.
7. Cole et al. (1996).
8. Lane and Wegner (1995).
9. Ibid.
10. Burgess and Holmstrom (1974).
11. Pennebaker (1990).
12. Davis and Schwartz (1987); Myers and Brewin (1994); Myers et al. (1998).
13. Myers and Brewin (1994).
14. Weinberger et al. (1979); King et al. (1990); Asendorpf and Scherer (1983).
15. King et al. (1990).
16. Niaura et al. (1992); Jamner et al. (1988); Brown et al. (1996).
17. Jamner et al. (1988); Jamner and Leigh (1999); Esterling et al. (1993).
18. Gross (1989); Jensen (1987); Temoshok (1987).
19. Pennebaker (1990).
20. Pennebaker and O'Heeron (1984).
21. Ibid.
22. Nolen-Hoeksema et al. (1993, 1997).

CHAPTER 5: EMOTIONAL DISCLOSURE:
THE REMARKABLE BENEFITS OF OPENING UP

1. Interview with Matt Varney (2001).
2. Pennebaker (1990).
3. Ibid.
4. Landwirth (1996).
5. Pennebaker (1990).
6. Ibid.

7. Pennebaker, Barger, Tiebout (1989).
8. Greenberg and Stone (1992); Greenberg, Wortman, Stone (1996); Spera et al. (1994).
9. Pennebaker and Beall (1986); Kelly et al. (1997); Greenberg and Stone (1992); Greenberg, Wortman, Stone (1996); Francis and Pennebaker (1992); Pennebaker (1997).
10. Pennebaker and Beall (1986); Francis and Pennebaker (1992); Pennebaker, Barger, Tiebout (1989).
11. Smyth et al. (1999).
12. Esterling et al. (1994); Lutgendorf et al. (1994); Booth et al. (1997); Petrie et al. (1995); Christensen et al. (1996); Pennebaker, Kiecolt-Glaser, Glaser (1988).
13. Petrie et al. (1998, 1995).
14. Spera et al. (1994).
15. Pennebaker (1990).
16. Pennebaker and Francis (1996); Pennebaker (1990).
17. Interview with Margie Levine (2001).
18. Levine (2001).
19. Interview with Louise DeSalvo, professor of English at Hunter College in New York and the author of *Writing as a Way of Healing: How Telling Our Stories Transforms Our Lives.* DeSalvo is a literary scholar, biographer, and memoirist whose other works include *Breathless, An Asthma Journal, Vertigo,* and *Adultery.*
20. Pennebaker (1990).
21. Greenberg et al. (1996).
22. Pennebaker et al. (1997); Pennebaker (1993).
23. Smyth et al. (2001).
24. Interview with Walter Anderson.
25. Kumin (2000).
26. With permission from Terry McMillan.
27. Pennebaker (1989).
28. DeSalvo (1999).
29. Pennebaker (1990).
30. Ibid.
31. Pennebaker et al. (1987); Lutgendorf et al. (1994); Esterling et al. (1990).
32. Pennebaker (1990).
33. Interview with Jonathan Progoff (2001), son of Ira Progoff, who created the intensive journaling method. J. Progoff is the director of Dialogue House, Inc. (NYC), national heaquarters for the Intensive Journal program. For more information visit www.intensivejournal.org or call 212-673-5880.

PART III: ENVIRONMENT AND RELATIONSHIPS: SOCIAL IMMUNITY

CHAPTER 6: HEALING BONDS

1. Interview with Lisa Berkman (2001).
2. Cassel (1976).
3. Berkman and Syme (1979).
4. Seeman et al. (1987).
5. Russek and Schwartz (1999).
6 Russek and Schwartz (1997b).
7. House, Robbins, Metzner (1982).
8. Blazer (1982).
9. Orth-Gomer, J. Johnson (1987).
10. Kaplan et al. (1988).
11. Berkman, Leo-Summers, Horwitz (1992).
12. Williams et al. (1992).
13. Cohen et al. (1997).
14. Leserman et al. (2000).
15. Feldman et al. (2000).
16. *The Providence Journal* (2001); Robert Sullivan wrote a compelling profile of Diana Golden Brosnihan for *Life* magazine in the summer of 1997, "Love Is a Reason to Live." He did a follow-up story on August 31, 2001, a few days after Brosnihan died, which can be found on Time.com.
17. Gove (1973); Hu and Goldman (1990).
18. Goodwin et al. (1987).
19. Waldron, Hughes, Brooks (1996); Goldman (1993).
20. Waite and Gallagher (2001).
21. Krzyzewski and Donald T. Phillips (2001).
22. Delany, Delany, Hearth (1996).
23. For more information, contact Share the Care, Murray Hill Station, P.O. Box 1217, New York, New York 10156. You can obtain a copy of "Share the Care: How to Organize a Group to Care for Someone Who Is Seriously Ill" written by Sheila Warnock and Cappy Caposella.
24. Cunningham (1997).
25. Albom (1997).
26. Cohen and Wills (1985); Cohen (1988); Broman (1993).
27. Uchino, Uno, Holt-Lunstad (1999); Uchino, Cacioppo, Kiecolt-Glaser (1996).
28. Strogatz et al. (1997).
29. Orth-Gomer Horsten et al. (1998).

30. Kiecolt-Glaser et al. (1991).
31. Theorell et al. (1995).
32. Seeman et al. (1994).
33. Lepore (1998).

CHAPTER 7: MIXED BLESSINGS:
THE COMPLEXITY OF SOCIAL RELATIONSHIPS

1. Ornish (1999).
2. Flay, Brian R. (1985); Gritz (1984).
3. Burg and Seeman (1994).
4. Seeman, Bruce, McAvay (1996).
5. Friedman et al. (1995).
6. Antonucci and Akiyama (1987).
7. Kiecolt-Glaser et al. (1997); Malarkey et al. (1994).
8. Putnam (2001).
9. Ibid.
10. Kawachi et al. (1997).
11. Kawachi, Kennedy, Glass (1999).
12. Ibid.
13. Fawzy et al. (1993); Fawzy et al. (1990); Spiegel, Bloom, Yalom (1981).

PART IV: PERSONAL ACHIEVEMENT AND EQUALITY:
LEARNING, EARNING, AND SURVIVING

1. Davey Smith et al. (1992).
2. Miringoff, Miringoff, Opdycke (2001).

CHAPTER 8: BEYOND OBELISKS: THE MYSTERY OF THE GRADIENT

1. *Webster's New World Dictionary and Thesaurus* (1998).
2. It is not clear at this time whether the effects of SEP are similar in developing countries as in industrial nations. See for example, Bunker et al. (1992).
3. Marmot, Shipley, Rose (1984); Rogot et al. (1992); Kaplan et al. (1996); Adler et al. (1993); Kitagawa and Hauser (1973); Feldman et al. (1989); Comstock, Tonascia (1978); Keil et al. (1984); Keil et al. (1992); Adler et al. (1994); Pincus, Callahan, Burkhauser (1987); Kaplan and Keil (1993); Marmot, Kogevinas, Elston (1987).
4. Williams et al. (1992); Ruberman et al. (1984).

5. Marmot and Shipley (1996); Rose and Marmot (1981).
6. Marmot, Bobak, Davey Smith (1995); Rose and Marmot (1981); Marmot, Shipley, Rose (1984).
7. Marmot and Shipley (1996).
8. Marmot, Shipley, Rose (1984).
9. Adler et al. (1993).
10. Rogot et al. (1992).
11. Adler et al. (1993); Adler et al. (1994).
12. Adapted from Pincus et al. (1987).
13. Williams et al. (1992).
14. Wilson et al. (1993); Dyer et al. (1976); Matthews et al. (1989); Gump, Matthews, Raikkonen (1999); Brunner (1997); Brunner et al. (1997); Schechter et al. (1994); Kubzansky, Kawachi, Sparrow (1999).
15. Gallo et al. (2001).
16. Brunner et al. (1997).
17. Lantz et al. (1998); Lynch, Kaplan, Salonen (1997); Winkleby, Fortmann, Barrett (1990); Kubzansky et al. (1998); Edmonds et al. (2001).
18. Marmot and Davey Smith (1997).
19. Story, Neumark, and French (2002).
20. Wilcox (1991); Altman, Schooler, Basil (1991).
21. Smith (1999).
22. Miringoff et al. (2001).
23. Baum, Garofalo, Yali (1999).
24. Redelmeier and Singh (2001).
25. Ibid.
26. Olshansky, Carnes, Cassel (1990); Desky and Redelmeier (1998).
27. Interview with Nancy Adler (2001).
28. Psychologists have long studied something called social comparison theory—the idea that we evaluate ourselves and our achievements partly by looking at how we compare with others. It is as if other people serve as a mirror through which we see our achievements more clearly. Depending on the issue, we compare ourselves with those who are doing less well (downward comparisons) or with those who are better off (upward comparisons). See Ramachandran (1994).
29. Manuck et al. (1995).
30. Pickering et al. (1988).
31. Kleinke and Williams (1994).
32. Adler et al. (2000).
33. Ibid.
34. Ostrove et al. (2000).

35. Cohen, Kaplan, Salonen (1999).
36. Seeman and Lewis (1995); Marmot et al. (1997).
37. Schnall and Landsbergis (1994).
38. According to Dr. Heymann, although the Family Medical Leave Act (FMLA) of 1993 requires employers to provide up to twelve weeks of unpaid leave to care for a child, parent, or spouse, it actually covers only half of all working adults. The reasons for this, among others, is that many work settings do not meet eligibility requirements (fifty or more employees), the employee has not been employed at the same work setting long enough, and the FMLA covers only births, adoptions, and *major* illnesses, of which most children's chronic illnesses do not qualify (Heymann, 2000).
39. Rodin and Langer (1977); Langer and Rodin (1976).
40. Starfield (1982).
41. Kaplan and Salonen (1990).
42. Power, Manor, Fox (1991); Wadsworth (1991); Wadsworth (1997).
43. Barker (1992, 1995).
44. Singer and Ryff (1999).
45. McEwen (1998); McEwen and Stellar (1993).

CHAPTER 9: BEYOND INDIVIDUAL ACHIEVEMENT: INEQUALITY AND RACE

1. Reich (2001).
2. Wilkenson (1992).
3. Kennedy et al. (1996).
4. Wilkenson (1996).
5. Wilkenson (1992).
6. Kawachi et al. (1997).
7. The terms "black" and "African American" are used interchangeably. The U.S. Office of Management and Budget (OMB) codes race in five categories: white, black, American Indian or Alaska Native, Asian or Pacific Islander, and Other. Ethnicity is defined as either Hispanic or Latino pertaining to a person of Cuban, Mexican, Puerto Rican, South or Central American, or other Spanish culture or origin, regardless of race.
8. Jaynes (1989).
9. Williams (2001); Kington and Nickens (2001).
10. National Human Genome Research Institute (2001); Lewontin (1973). The lack of substantial differences in genetic structure between racial groups does not imply that genes are unimportant in the high prevalence

of illnesses in African Americans. Genes likely interact with other behavioral, psychological, or social factors to affect health outcomes in this group. In fact, psychosocial and behavioral factors might lead to racial group differences in *gene expression* even if the structure of DNA is similar.

11. Oliver and Shapiro (2001).
12. Otten et al. (1990).
13. Pappas et al. (1993).
14. Wilson (1987).
15. Haan et al. (1987).
16. Auberbach et al. (2000).

PART V: FAITH AND MEANING: EXISTENTIAL, RELIGIOUS, AND SPIRITUAL DIMENSIONS OF HEALTH

1. Frankl (1959, 1962, 1984).
2. Helmrich (1992).

CHAPTER 10: FROM TRAUMA TO MEANING

1. Levine (2001).
2. Ibid.
3. Parkes (1975).
4. Janoff-Bulman (1989); Fiske and Linville (1980); Fiske and Taylor (1984); Hastie (1981).
5. Janoff-Bulman (1992); Anderson, Lepper, Ross (1980); Swann and Read (1981); Rothbart, Evans, Fulero (1979); Cantor and Michsel (1979); Langer and Abelson (1974).
6. Janoff-Bulman (1992).
7. Janoff-Bulman (1989).
8. Ibid.
9. Lerner (1980).
10. Janoff-Bulman (1989).
11. Levine (2001).
12. For an overview of the concept of finding meaning in or growing from adversity, see Tedeschi, Park, Calhoun (1998); Park and Folkman (1997); Park et al. (1996).
13. Price (2000).
14. Schafer and Moss (1992).
15. Interview with Representative Tom Osborn (2001).
16. Greenberg (1995); Taylor (1983).

17. Taylor (1983).
18. Welch-McCaffrey et al. (1989).
19. Thompson (1991).
20. Fromm, Andrykowski, Hunt (1996).
21. Schwartzberg (1994).
22. Siegel and Schrimshaw (2000).
23. Affleck, Tennen, Rowe (1991).
24. Coffman (1994); Saylor, Swenson, Powell (1992).
25. Nelson (1989, 1994).

CHAPTER 11: THE HEALTH BENEFITS OF FINDING MEANING

1. Aldwin, Levenson, Spiro III (1994).
2. Davis, Nolen-Hoeksema, Larson (1998).
3. Cruess et al. (2000).
4. Affleck et al. (1987).
5. Bower et al. (1998).
6. Affleck et al. (1991).

CHAPTER 12: FAITH, MEANING, AND LONGEVITY

1. Jones (1994).
2. Freud (1927, 1962).
3. Gallup (1995).
4. The results of this evaluation are currently under review for publication.
5. Interview with Harold Koenig (2001).
6. McCullough et al. (2000).
7. Strawbridge et al. (1997).
8. Koenig (1999).
9. Koenig et al. (1998); Larson et al. (1989).
10. Ellison (1990); Levin, Chatters, Taylor (1995); Koenig, Kvale, Ferrel (1988); Levin, Markides, Ray (1988); Koenig et al. (1992); Koenig, George, Peterson (1998).
11. Woods et al. (1999).
12. Koenig et al. (1997).
13. Koenig and Larson (1998).
14. Hardestym and Kirby (1995); Alexander and Duff (1991); Kendler, Gardner, Prescott (1997); Koenig et al. (1998).
15. Interview with Maya Angelou (2001).
16. Pargament (1997).

17. McIntosh (1995).
18. Helmrich (1992).
19. Koenig (1999).
20. Ibid.
21. Pargament (1997).
22. Maton (1989).
23. Krause and Van Tran (1998).
24. Ibid.
25. Tix and Frazier (1998).
26. Ibid.
27. Koenig et al. (1992).
28. Tix and Frazier (1998).
29. Steffen et al. (2001).
30. Ellison (1993); Krause and Van Tran (1989); Lincoln and Mamiya (1990).
31. Krause (1998).
32. Argyle (1994); Polner (1989).
33. Pargament (1997).
34. Oxan and Reed (1998).
35. Interview with Andrew Young (2001).
36. McIntosh, Silver, Wortman (1993).
37. SIDS fact sheet, NICHD, NIH. Online at http://www.nichd.nih.gov/publications/pubs/sidsfact.htm.
38. Fennell's story was told in the *Washington Post,* June 19, 1999, Section E, page 1, Warren Brown.
39. Lisa Chick and her children were interviewed November 19, 2001 by Robin Roberts on *Good Morning America* as a part of their Recovery 101 series.
40. The Terry Fox Foundation; www.terryfoxrun.org; 888-836-9786.
41. Hendrick (2001).
42. CBS correspondent Harold Dow reported on Tim Streett for *48 Hours,* December 5, 2001.
43. Brown (2001).

PART VI: CONNECTIONS—TO EMOTIONS, TO THE FUTURE

CHAPTER 13: EMOTIONAL CONNECTIONS

1. Lazarus and Lazarus (1994).
2. LeDoux (1996).

3. Lazarus (1994).
4. *Diagnostic and Statistical Manual of Mental Disorders,* 4th ed. (1994).
5. Kubzansky and Kawachi (2000); Musselman, Evans, Nemeroff (1998); Rozanski, Blumenthal, Kaplan (1999).
6. Kubzansky et al. (1998); Rozanski et al. (1999).
7. Koskenvuo et al. (1988); Kawachi et al. (1996).
8. Kubzansky et al. (1998); Gallagher et al. (1999).
9. Miller et al. (1996).
10. Smith (1992).
11. Miller et al. (1996).
12. Barefoot, Dahlstrom, Williams (1983).
13. Lazarus and Folkman (1984).
14. New Testament, Romans 8:28.
15. Golenbeck (1990).
16. Walker (2001); Corkery (2001); Milkovits (2001); accounts of Sergeant Young's shooting and the events that followed appeared in papers across the United States. Facts here are from articles in three New England newspapers.
17. Engel (1971); Hafen et al. (1996).
18. Rosengren et al. (1993).
19. Ruberman et al. (1984).
20. Leserman et al. (2000); Leserman et al. (1999).
21. Cohen, Tyrrell, Smith (1991).
22. The effects of stress were not altered after taking into account, age, sex, education, allergic status, weight, the season of the year, number of participants quarantined together, or antibody status before virus exposure. The findings were also not explained by smoking, alcohol consumption, exercise, diet, sleep quality, white blood cells, or immunoglobin levels.
23. Bosma and Marmot (1997); Karasek et al. (1988); Schnall et al. (1994); Schnall et al. (1990).
24. Martikainen and Valkonen (1996).
25. Ann Bowling (1987).
26. Ory et al. (1999).
27. Schulz et al. (1997); Vitaliano et al. (1991).
28. Schulz and Beach (1999).

CHAPTER 14: EMOTIONS AS "CONNECTIVE TISSUE"

1. Seligman (1990).
2. Cohen and Wills (1985).

3. Rosengren et al. (1993).
4. Gallo and Matthews (1997).
5. Baum et al. (1997).
6. Anderson (1998).
7. Baum and Posluszny (1999).
8. Jiang et al. (1996).
9. Blumenthal et al. (1997).
10. Kaplan (1995).
11. Ruberman et al. (1984).

CHAPTER 15: CONNECTING TO THE FUTURE

1. Fredrickson and Levenson (1998); Fredrickson (2000); Fredrickson (1998); Fredrickson and Joiner (2002).
2. Masten (2001).
3. Silver et al. (2002).
4. Surwit et al. (1984); Lehman et al. (1991); Bank (1988); Bank, LoTurco, Alkon (1989); Kuhn and Schanberg (1998); Meany et al. (2000).
5. Hong et al. (1999); Glaser et al. (1993).
6. Olds et al. (1997); Field (1995); Lorig et al. (1998); Wing et al. (2001); McQuaid and Nassau (1999); Knowler et al. (2002); Smith, Kendall, Keefe (2002); Fawzy et al. (1993).
7. Kaplan and Groessl (2002); Blumenthal et al. (2002).
8. Kington and Nickens (2001).
9. Addressing Health Disparities: The NIH Program of Action, National Institutes of Health, www.healthdisparities.hi.gov.
10. National Institute of Health, "Toward Higher Levels of Analysis: Progress and Promise in Research on Social and Cultural Dimensions of Health," executive summary, Office of Behavioral or Social Sciences Research, NIH Publication No. 21-5020, September 2001, www.obsst.od.gov/publications.

Bibliography

Adler, N.E. et al. (1993). "Socioeconomic Inequalities in Health: No Easy Solution," *JAMA,* Vol. 269, pp. 3140–45.

——— et al.(1994). "Socioeconomic Status and Health: The Challenge of the Gradient," *American Psychologist,* Vol. X, January, pp. 1–10.

——— and E. S. Epel (2000). "Relationship of Subjective and Objective Social Status with Psychological and Physiological Functioning: Preliminary Data in Healthy White Women," *Health Psychology,* Vol. 19, pp. 586–92.

Affleck, G. et al. (1987). "Causal Attribution, Perceived Benefits, and Morbidity After a Heart Attack: An 8-Year Study," *Journal of Consulting and Clinical Psychology,* Vol. 55, pp. 29–35.

Afflect, G., H. Tennen, and J. Rowe (1991). *Infants in Crisis: How Parents Cope with Newborn Intensive Care and Its Aftermath.* New York: Springer-Verlag.

Albom, Mitch (1997). *Tuesdays with Morrie: An Old Man, a Young Man, and Life's Greatest Lesson.* New York: Doubleday.

Aldwin, C.M., M. R. Levenson, and A. Spiro III (1994). "Vulnerability and Resilience to Combat Exposure: Can Stress Have Lifelong Effects?," *Psychology and Aging,* Vol. 9, pp. 34–44.

Alexander, F. and R. W. Duff (1991). "Influence of Religiosity and Alcohol Use on Personal Well-being," *Journal of Religious Gerontology,* Vol. 8, pp. 11–21.

Alloy, L.B. and L. Y. Abramson (1979). "Judgement of Contingency in Depressed and Nondepressed Students: Sadder but Wiser?," *Journal of Experimental Psychology,* Vol. 108, pp. 441–85.

Altman, D.G., C. Schooler, and M. D. Basil (1991). "Alcohol and Cigarette Advertising on Billboards," *Health Education Research,* Vol. 6, pp. 487–90.

Anderson, C., M. R. Lepper, and L. Ross (1990). "Perseverance of Social Theories: The Role of Explanation in the Persistence of Discredited Information," *Journal of Personality and Social Psychology,* Vol. 39, pp. 1037–49.

Anderson, N.B. (1998). "Levels of Analysis in Health Science: A Framework for Integrating Sociobehavioral and Biomedical Research," *Annals of the New York Academy of Sciences*, Vol. 840, pp. 563–76.

Antonucci, T.C. and H. Akiyama (1987). "An Examination of Sex Differences in Social Support Among Older Men and Women," *Sex Roles*, Vol. 17, pp. 737–49.

Argyle, M. (1994). *The Psychology of Social Class*. New York: Routledge.

Asendorph, J.B. and K. R. Scherer (1983). "The Discrepant Repressor: Differentiation Between Low Anxiety, High Anxiety, and Repression of Anxiety by Autonomic-Facial-Verbal Patterns of Behavior," *Journal of Personality and Social Psychology*, Vol. 45, pp. 1334–346.

Auerbach, Krimgold, and Lefkowitz (2000). "Improving Health: It Doesn't Take a Revolution." National Policy (Association Report #298). Washington, D.C.: National Policy Association.

Bandura, A. (1997). *Self-Efficacy: The Exercise of Control*, Freeman, p. 11.

————— et al. (1985). "Catecholamine Secretion as a Function of Perceived Coping Self-Efficacy," *Journal of Consulting and Clinical Psychology*, Vol. 53, pp. 406–14.

————— et al. (1999). "Self-Efficacy Pathways to Childhood Depression," *Journal of Personality and Social Psychology*, Vol. 76, pp. 258–69.

————— et al. (1987). "Perceived Self-Efficacy and Pain Control: Opioid and Nonopioid Mechanisms," *Journal of Personality and Social Psychology*, Vol. 53, pp. 563–71.

—————, N. E. Adams, and J. Breyer (1977). "Cognitive Processes Mediating Behavioral Change," *Journal of Personality and Social Psychology*, Vol. 35, p. 125.

—————, L. Reese, and N. E. Adams (1982). "Microanalysis of Action and Fear Arousal as a Function of Differential Levels of Perceived Self-Efficacy," *Journal of Personality and Social Psychology*, Vol. 43, pp. 5–21.

Bank, B. et al. (1988). "Classical Conditioning Induces Long-Term Translocation of Protein Kinase C in Rabbit Hippocampal CA1 Cells," *Proc. Natl. Acad. Sci.*, Vol. 85, pp. 1988–92.

—————, J. J. LoTurco, and D. L. Alkon (1989). "Learning-Induced Activation of Protein Kinase C: A Molecular Memory Trace," *Mol. Neurobiol*, Vol. 3, pp. 55–70.

Barefoot, J.C., W. G. Dahlstrom, and R. B. Williams (1983). "Hostility, CHD Incidence and Total Mortality: A 25-Year Follow-up Study of 255 Physicians," *Psychosom Med.*, Vol. 45, pp. 59–63.

Barker, D. (1992). "Fetal and Infant Origins of Adult Disease," *BMJ*, Vol. 25.

————— (1995). "Fetal Origins of Coronary Heart Disease," *BMJ*, Vol. 31, pp. 171–74.

Barras, J.R. (2000). *Whatever Happened to Daddy's Little Girl: The Impact of Fatherlessness on Black Women*. New York: One World.

Baum, A., J. Garofalo, and A. Yali (1999). "Socioeconomic Status and Chronic Stress: Does Stress Account for SES Effects on Health?" In N. Adler et al., "Socioeconomic Status and Health in Industrial Nations: Social, Psychological, and Biological Pathways," *Annals of the New York Academy of Sciences*, Vol. 896, pp. 131–44.

——— and D. Posluszny (1999). "Health Psychology: Mapping Biobehavioral Contributions to Health and Illness." *Annual Review of Psychology*, Vol. 50, pp. 137–63.

Beevers, C.G., R. M. Wenzlaff, and A. M. Hayes (1999). "Depression and the Ironic Effects of Thought Suppression: Therapeutic Strategies for Improving Mental Control," *Clinical Psychology: Science and Practice*, Vol. 6, pp. 133–48.

Berkman, L.F., and S. L. Syme (1979). "Social Networks, Host Resistance and Mortality: A Nine-Year Follow-up Study of Alameda County Residents," *Am. J. Epidemiol*, Vol. 109, pp. 186–204.

———, L. Leo-Summers, and R. I. Horwitz (1992). "Emotional Support and Survival Following Myocardial Infarction: A Prospective Population-Based Study of the Elderly," *Ann Intern Med*, Vol. 117, pp. 1003–9.

Blazer, D. (1982). "Social Support and Mortality in an Elderly Community Population," *Am J Epidemiol*, Vol. 115, pp. 684–94.

Blumenthal, J.A. et al. (1997). "Stress Management and Exercise Training in Cardiac Patients with Myocardial Ischemia: Effects on Prognosis and Evaluation of Mechanisms," *Archives of Internal Medicine*, Vol. 157, pp. 2213–23.

——— et al. (2002). "Usefulness of Psychosocial Treatment of Mental Stress-Induced Myocardial Ischemia in Men," *Am J Cardiol*, Vol. 89, pp. 164–8.

Bone, C.R. et al. (1992). "Research Needs and Opportunities Related to Respiratory Health of Women," *American Review of Respiratory Diseases*, pp. 328–525.

Booth, R.J. et al. (1997). "Changes in Circulating Lymphocyte Numbers Following Emotional Disclosure: Evidence of Buffering?" *Stress Medicine*, Vol. 13, pp. 23–29.

——— and J. Pennebaker (1998). "The Immunological Effects of Thought Suppression," *Journal of Personality and Social Psychology*, Vol. 75, pp. 1264–72.

Bosma, H. et al. (1997) "Low Job Control and Risk of Coronary Heart Disease in Whitehall II Study," *BMJ*, Vol. 314, pp. 558–65.

Bower, J.E. et al. (1998). "Cognitive Processing, Discovery of Meaning, CD4 Decline, and AIDS-Related Mortality Among Bereaved HIV-Seropositive Men," *Journal of Consulting and Clinical Psychology*, Vol. 66, pp. 979–86.

Bowling, A. (1987). "Mortality After Bereavement: A Review of the Literature on Survival Periods and Factors Affecting Survival," *Soc. Sci. Med.*, Vol. 24, pp. 117–24.

Broman, C.L. (1993). "Social Relationships and Health-Related Behavior," *Journal of Behavioral Medicine*, Vol. 16, pp. 335–50.

Bromberger, J.R. and K. A. Matthews (1996). "A Longitudinal Study of the Effects of Pessimism, Trait Anxiety, and Life Stress on Depressive Symptoms in Middle-Aged Women," *Psychology and Aging*, Vol. 11, pp. 207–13.

Brown, L.L. et al. (1996). "Individual Differences in Repressive-Defensiveness Predict Basal Salivary Cortisol Levels," *Journal of Personality and Social Psychology*, Vol. 70, pp. 362–71.

Brown, Z. "A Wake-Up Call to Greater Good," *Unity Magazine*, October 2001, pp. 7–12.

Brunner, E. (1997). "Stress and the Biology of Inequality," *BMJ*, Vol. 314, pp. 1472–76.

—— et al. (1997). "Social Inequality in Coronary Risk: Central Obesity and the Metabolic Syndrome: Evidence from the Whitehall II Study," *Diabetologia*, Vol. 40, pp. 1341–49.

Bunker, C.H. et al. (1992). "Factors Associated with Hypertension in Nigerian Civil Servants," *Prev Med*, Vol. 21, pp. 710–22.

Burg, M.M. and T. E. Seeman (1994). "Families and Health: The Negative Side of Social Ties," *Annals of Behavioral Medicine*, Vol. 16, pp. 109–15.

Burgess, A.W. and L. L. Holmstrom (1974). "Rape Trauma Syndrome," *American Journal of Psychiatry*, Vol. 131, pp. 981–86.

Cacioppo, J.T. (1994). "Social Neuroscience: Autonomic, Neuroendocrine, and Immune Responses to Stress," *Psychophysiology*, Vol. 31, pp. 113–28.

—— (2000). "Autonomic, Neuroendocrine, and Immune Responses to Psychological Stress," *Psychologische Beitrage*, Vol. 42, pp. 4–23.

Cantor, N. (1980). "Perceptions of Situation: Situation Prototypes and Person-Situation Prototypes." In D. Magnusson, ed., *The Situation: An Interactional Perspective*. Hillsdale, N.J.: Erlbaum.

—— and W. Mischel (1979). "Prototypes in Person Perception." In L. Berkowitz, ed., *Advances in Experimental Social Psychology*, Vol. 12. New York: Academic Press.

Carver, C.S. and J. Gollin Gaines (1987). "Optimism, Pessimism, and Postpartum Depression," *Cognitive Therapy and Research*, Vol. 11, pp. 449–62.

—— and M. F. Scheier (1999). "Optimism" in C. R. Snyder, ed., *Coping: The Psychology of What Works*. Oxford: Oxford University Press, pp. 3–19.

——— et al. (1993). "How Coping Mediates the Effect of Optimism on Distress: A Study of Women with Early Stage Breast Cancer," *Journal of Personality and Social Psychology,* Vol. 65, pp. 375–90.

Cassel, J. (1976). "The Contribution of the Social Environment to Host Resistance," *American Journal of Epidemiology,* Vol. 104, p. 107.

Christensen, A.J. et al. (1996). "Effect of Verbal Self-Disclosure on Natural Killer Cell Activity: Moderating Influence of Cynical Hostility," *Psychosomatic Medicine,* Vol. 58, pp. 150–55.

Clark, N.M. and A. Dodge (1999). "Exploring Self-Efficacy as a Predictor of Disease Management," *Health Education & Behavior,* Vol. 26, pp. 72–89.

Coffman, S. (1994). "Children Describe Life After Hurricane Andrew," *Pediatric Nursing,* Vol. 20, pp. 363–75.

Cohen, S. (1988). "Psychosocial Models of the Role of Social Support in the Etiology of Physical Disease," *Health Psychology,* Vol. 7, pp. 269–97.

———, D. Tyrrell, and A. P. Smith (1991). "Psychological Stress and Susceptibility to the Common Cold," *N Engl J Med,* Vol. 325, pp. 606–12.

——— and T. A. Wills (1985). "Stress, Social Support, and the Buffering Hypothesis," *Psychological Bulletin,* Vol. 98, pp. 310–57.

——— et al. (1997). "Social Ties and Susceptibility to the Common Cold," *Journal of the American Medical Association,* Vol. 277, pp. 1940–44.

——— and J. T. Salonen (1999). "The Role of Psychological Characteristics in the Relation Between Socioeconomic Status and Perceived Health," *Journal of Applied Social Psychology,* Vol. 29, pp. 445–68.

Cole, S. et al. (1996). "Accelerated Course of Human Immunodeficiency Virus Infection in Gay Men Who Conceal Their Homosexual Identity," *Psychosomatic Medicine,* Vol. 58, pp. 219–31.

Comstock, G.W. and J. A. Tonascia (1978). "Education and Mortality in Washington County, Maryland," *J Health Soc Behav.,* Vol. 18, pp. 54–61.

Coyne, J.C. and I. H. Gotlieb (1983). "The Role of Cognition in Depression: A Critical Appraisal," *Psychological Bulletin,* pp. 472–505.

Cruess, D.G. et al. (2000). "Cognitive-Behavioral Stress Management Reduces Serum Cortisol by Enhancing Benefit Finding Among Women Being Treated for Early Stage Breast Cancer," *Psychosomatic Medicine,* Vol. 22, pp. 304–8.

Davey Smith, G. et al. (1992). "Socio-economic Differentials in Mortality: Evidence from Glasgow Graveyards," *BMJ,* Vol. 305, pp. 1554–57.

Davis, C.G., S. Nolen-Hoeksema, and J. Larson (1998). "Making Sense of Loss and Benefiting from the Experience: Two Construals of Meaning," *Journal of Personality and Social Psychology,* Vol. 75, pp. 561–74.

Davis, P.J. and G. E. Schwartz (1987). "Repression and the Inaccessibility of Affective Memories," *Journal of Personality and Social Psychology*, Vol. 52, pp. 155–62.

Delany, S.L., Delany, A. E., and A. Hearth, ed. (1996). *Having Our Say: The Delany Sisters' First 100 Years*. New York: Dell.

Detsky, A.S. and D. A. Redelmeier (1998). "Measuring Health Outcomes—Putting Gains into Perspective," *N Engl J Med.*, Vol. 339, pp. 402–4.

Diagnostic and Statistical Manual of Mental Disorders—Fourth Edition (2000). Washington, D.C.: The American Psychiatric Association.

Dolce, J.J. (1987). "Self-Efficacy and Disability Beliefs in Behavioral Treatment of Pain," *Behaviour Research and Therapy*, Vol. 25, pp. 289–99.

Dominguez, B. et al. (1995). "The Roles of Emotional Reversal and Disclosure in Clinical Practice," in J. W. Pennebaker, ed., *Emotion, Disclosure, and Health*. Washington, D.C.: American Psychological Association.

Dyer, A.R. et al. (1976). "The Relationship of Education to Blood Pressure: Findings on 40,000 Employed Chicagoans," *Circulation*, Vol. 54, pp. 987–92.

Edmonds, J. et al. (2001). "Ecological and Socioeconomic Correlates of Fruit, Juice, and Vegetable Consumption Among African-American Boys," *Prev Med*, Vol. 32, pp. 476–81.

Ellison, C.G. (1991). "Religious Involvement and Subjective Well-being," *Journal of Health and Social Behavior*, Vol. 32, p. 90.

———— (1993). "Religious Involvement and Self-Perception Among Black Americans," *Social Forces*, pp. 1027–55.

Engel, G.L. (1971). "Sudden Rapid Death During Psychological Stress: Folklore or Folk Wisdom?" *Annals of Internal Medicine*, Vol. 74, pp. 771–82.

Esterling, B.A. et al. (1990). "Emotional Repression, Stress Disclosure Responses, and Epstein-Barr Viral Capsid Antigen Titers," *Psychosomatic Medicine*, Vol. 52, pp. 397–410.

———— et al. (1993). "Defensiveness, Trait Anxiety, and Epstein-Barr Viral Capsid Antigen Antibody Titers in Healthy College Students," *Health Psychology*, Vol. 12, pp. 132–39.

———— et al. (1994). "Emotional Disclosure Through Writing or Speaking Modulates Latent Epstein-Barr Virus Reactivation," *Journal of Consulting and Clinical Psychology*, Vol. 62, pp. 130–40.

Fawzy, F.I. et al. (1990). "A Structured Psychiatric Intervention for Cancer Patients: II. Changes over Time in Immunological Measures," *Archives of General Psychiatry*, Vol. 47, pp. 729–35.

———— et al. (1993). "Malignant Melanoma: Effects of an Early Structured Psychiatric Intervention, Coping, and Affective State on Recurrence and Survival 6 Years Later," *Archives of General Psychiatry*, Vol. 50, pp. 681–89.

Feldman, J.J. et al. (1989). "National Trends in Educational Differentials in Mortality," *Am J Epidemiol*, Vol. 129, pp. 919–33.

Feldman, P.J. et al. (2000). "Material Social Support Predicts Birth Weight and Fetal Growth in Human Pregnancy," *Psychosomatic Medicine*, Vol. 62, pp. 715–25.

Field, T. (1995). "Massage Therapy for Infants and Children," *Journal of Developmental & Behavioral Pediatrics*, Vol. 16, pp. 105–11.

Fiske, S. and P. Linville (1980). "What Does the Schema Concept Buy Us?," *Personality and Social Psychology Bulletin*, Vol. 6, pp. 543–57.

——— and S. Taylor (1984). *Social Cognition*. Boston, Mass.: Addison-Wesley.

Fitzgerald, T.E. et al. (1993). "The Relative Importance of Dispositional Optimism and Control Appraisals in Quality of Life After Coronary Artery Bypass Surgery," *Journal of Behavioral Medicine*, Vol. 16, pp. 24–43.

Flay, B.R. (1985). "Adolescent Smoking: Onset and Prevention," *Annals of Behavioral Medicine*, Vol. 7, pp. 9–13.

Fontain, K.R., A.S.R. Manstead, and H. Wagner (1993). "Optimism, Perceived Control over Stress, and Coping," *European Journal of Personality*, Vol. 7, pp. 267–81.

Francis, M.E. and J. W. Pennebaker (1992). "Putting Stress into Words: The Impact of Writing on Physiological, Absentee, and Self-Reported Emotional Well-Being Measures," *American Journal of Health Promotion*, Vol. 6, pp. 280–87.

Frankl, V.E. (1959, 1962, 1984). *Man's Search for Meaning*. New York: Simon & Schuster.

Fredrickson, B.L. (1998). "What Good Are Positive Emotions?" *Review of General Psychology*, Vol. 2, pp. 300–319.

——— (2000). "The Undoing Effect of Positive Emotions," *Motivation & Emotion*, Vol. 24, pp. 237–58.

——— and R. W. Levenson (1998). "Positive Emotions Speed Recovery from the Cardiovascular Sequelae of Negative Emotions," *Cognition & Emotion*, Vol. 12, pp. 191–220.

——— and T. Joiner (2002). "Positive Emotions Trigger Upward Spirals Toward Emotional Well-being," *Psychological Science*, Vol. 13, pp. 172–75.

Freidman, H.S. et al. (1995). "Psychosocial and Behavioral Predictors of Longevity: The Aging and Death of the 'Termites.'" *American Psychologist*, Vol. 50, pp. 69–78.

Fromm, K., M. A. Andrykowski, and J. Hunt (1996). "Positive and Negative Psychosocial Sequelae of Bone Marrow Transplantation: Implications for Quality of Life Assessment," *Journal of Behavioral Medicine*, Vol. 19, p. 221.

Fry, P.S. (1995). "Perfectionism, Humor and Optimism as Moderators of Health Outcomes and Determinants of Coping Styles of Women Executives," *Genetic, Social and General Psychology Monographs,* Vol. 121, pp. 211–45.

Gallacher, J.E.J. et al (1999). "Anger and Incident Heart Disease in the Caerphilly Study," *Psychosomatic Medicine,* Vol. 61, pp. 446–53.

Gallo, L. and K. Matthews (1999). "Do Emotions Mediate the Association Between Socioeconomic Status and Health?" In N. Adler et al., eds. "Socioeconomic Status and Health in Industrial Nations," *Annals of the New York Academy of Sciences,* Vol. 987, pp. 226–45.

———— et al. (2001). "Educational Attainment and Coronary and Aortic Calcification in Postmenopausal Women," *Psychosomatic Medicine,* Vol. 63, pp. 925–35.

Gerin, W. et al. (1995). "Self-Efficacy as a Moderator of Perceived Control Effects on Cardiovascular Reactivity: Is Enhanced Control Always Beneficial?," *Psychosomatic Medicine,* pp. 390–97.

Glanz, K. et al. (1995). "Environmental and Policy Approaches to Cardiovascular Disease Prevention Through Nutrition: Opportunities for State and Local Action," *Health Education Quarterly,* Vol. 22, pp. 512–27.

Glaser, R. et al. (1993). "Stress-Associated Modulation of Proto-Oncogene Expression in Human Peripheral Blood Leukocytes," *Behavioral Neuroscience,* Vol. 107, pp. 525–29.

Goldman, N. (1993). "Marriage Selection and Mortality Patterns: Inferences and Fallacies," *Demography,* Vol. 30, pp. 189–208.

Golenbeck, P. (1990). *Personal Fouls.* New York: Signet.

Goodwin, J.S. et al. (1987). "The Effect of Marital Status on Stage, Treatment, and Survival of Cancer Patients," *JAMA,* Vol. 258, pp. 3125–30.

Gove, W.R. (1973). "Sex, Marital Status, and Mortality," *Am. J. Sociol,* Vol. 79, pp. 45–67.

Grant, M.D., Z. H. Piotrowski and R. Chappell (1995)."Self-Reported Health and Survival in the Longitudinal Study of Aging, 1984–1986," *J Clin Epidemiol,* pp. 375–87.

Greenberg, M.A. (1995). "Cognitive Processing of Traumas: The Role of Intrusive Thoughts and Reappraisals," *Journal of Applied Social Psychology,* Vol. 25, pp. 1262–96.

———— and A. A. Stone (1992). "Emotional Disclosure About Traumas and Its Relation to Health: Effects of Previous Disclosure and Trauma Severity," *Journal of Personality and Social Psychology,* Vol. 63, pp. 75–84.

————, C. B. Wortman, and A. A. Stone (1996). "Emotional Expression and Physical Health: Revising Traumatic Memories or Fostering Self-Regulation?," *Journal of Personality and Social Psychology,* Vol. 71, pp. 588–602.

Greenberg, S. and K. Springen (2001). "The Baby Blues and Beyond," *Newsweek,* July 2, pp. 26–29.

Gritz, E.R. (1984). "Cigarette Smoking by Adolescent Females: Implications for Health and Behavior," *Women & Health,* Vol. 9, pp. 103–15.

Gross, J. (1989). "Emotional Expression in Cancer Onset and Progression," *Soc. Sci. Med.,* Vol. 28, pp. 1239–248.

Gump, B.B. and K. A. Matthews (1999). "Modeling Relationships Among Socioeconomic Status, Hostility, Cardiovascular Reactivity, and Left Ventricular Mass in African American and White Children," *Health Psychology,* Vol. 18, pp. 140–50.

Haan, M., G. Kaplan, and T. Camacho (1987). "Poverty and Health: Prospective Evidence from the Alameda County Study," *American Journal of Epidemiology,* Vol. 125, pp. 989–98.

Hafen, B., K. Karren, and N. Frandsen (1996). *Mind/Body Health: The Effects of Attitudes, Emotions, and Relationships.* Boston: Allyn & Bacon.

Hardestym, P.H. and K. M. Kirby (1995). "Relation Between Family Religiousness and Drug Use Within Adolescent Peer Groups," *Journal of Social Behavior and Personality,* Vol. 10, pp. 421–30.

Hastie, R. (1981). "Schematic Principles in Human Memory." In E. T. Higgins, C. P. Herman, and M. Zanna, eds., *Social Cognition.* Hillsdale, N.J.: Erlbaum.

Helmreich, W. (1992). *Against All Odds: Holocaust Survivors and the Successful Lives They Made in America.* New York: Transaction Publisher.

Hendrik, B. "Teen Driving: Honoring Rebekkah," *The Atlanta Constitution,* January 8, 2001, p. 1D.

Hensen, M.R. (1987). "Psychobiological Factors Predicting the Course of Breast Cancer," *Journal of Personality,* Vol. 55, p. 2.

Heymann, J. (2000). *The Widening Gap: Why America's Working Families Are in Jeopardy and What Can Be Done About It.* New York: Basic Books.

Hong, W. et al. (1999). "Chronic Stress Associated with Spousal Caregiving of Patients with Alzheimer's Dementia Is Associated with Downregulation of B-Lymphocite GH mRNA," *Journals of Gerontology: Series A: Biological Sciences & Medical Sciences,* Vol. 54, pp. M212–15.

Horowitz, M.J. (1975). "Intrusive and Repetitive Thoughts After Experiencing Stress," *Archives of General Psychiatry,* Vol. 32, pp. 1427–63.

—— (1986). *Stress Response Syndromes* (2nd ed.). New York: Jason Aronson.

House, J.S., C. Robbins, and H. L. Metzner (1982). "The Association of Social Relationships and Activities with Mortality: Prospective Evidence from the Tecumseh Community Health Study," *Am J Epidemiol,* Vol. 116, pp. 123–40.

Hu, Y. and N. Goldman (1990). "Mortality Differentials by Marital Status: An International Comparison," *Demography,* Vol. 27, pp. 233–50.

Idler, E.L. and R. J. Angel (1990). "Self-Rated Health and Mortality in the NHANES-I Epidemiologic Follow-up Study," *Am J Public Hlth,* pp. 446–52.

Jamner, L.D., G. E. Schwartz, and H. Leigh (1988). "The Relationship Between Repressive and Defensive Coping Styles and Monocyte, Eosinophile, and Serum Glucose Levels: Support for the Opiod Peptide Hypothesis of Repression," *Psychosomatic Medicine,* Vol. 50, pp. 567–75.

———— and H. Leigh (1999). "Repressive/Defensive Coping, Endogenous Opioids and Health: How a Life So Perfect Can Make You Sick," *Psychiatry Research,* Vol. 85, pp. 17–31.

Janoff-Bulman, R. (1989). "Assumptive Worlds and the Stress of Traumatic Events: Applications of the Schema Construct," *Social Cognition,* Vol. 7, pp. 113–36.

———— (1992). *Shattered Assumptions: Towards a New Psychology Trauma.* New York: The Free Press.

Jaynes, G., and R. Williams, Jr. (1989). *A Common Destiny: Blacks and American Society.* Washington, D.C.: National Academy Press.

Jensen, M.R. (1987). "Psychobiological Factors Predicting the Course of Breast Cancer: Personality and Physical Health," *J Pers* (Special Issue), Vol. 55, pp. 317–42.

Jiang, W. et al. (1996). "Mental Stress-Induced Myocardial Ischemia and Cardiac Events," *Journal of the American Medical Association,* Vol. 275, pp. 1651–56.

Jones, S.L. (1994). "A Constructive Relationship for Religion with the Science and Profession of Psychology: Perhaps the Boldest Model Yet," *American Psychologist,* Vol. 49, pp. 184–99.

Kabat-Zinn, J. (1995). *Mindfulness Meditation: Cultivating the Wisdom of Your Body and Mind.* New York: Simon & Schuster Audio.

———— (1998). "Influence of a Mindfulness Meditation-Based Stress Reduction Intervention on Rates of Skin Clearing in Patients with Moderate to Severe Psoriasis Undergoing Phototherapy (UVB) and Photochemotherapy (PUVA)," *Psychosomatic Medicine,* Vol. 60, pp. 625–32.

———— et al. (1985). "The Clinical Use of Mindfulness Meditation for the Self-Regulation of Chronic Pain," *Journal of Behavioral Medicine,* Vol. 8, pp. 163–90.

Kamen-Siegel, L. et al. (1991). "Explanatory Style and Cell-Mediated Immunity in Elderly Men and Women," *Health Psychology,* Vol. 10, pp. 229–35.

Kaplan, G., V. Barell, and A. Lusky (1988). "Subjective State of Health and Survival in Elderly Adults," *J Gerontol*, pp. S114–S20.

Kaplan, G.A. (1995). "Where Do Shared Pathways Lead? Some Reflections on a Research Agenda," *Psychosom Med*, Vol. 57, pp. 208–12.

———— and T. Camacho (1983). "Perceived Health and Mortality: A Nine-Year Follow-up of the Human Population Laboratory Cohort," *Am J Epidemiol*, pp. 895–904.

———— and J. T. Salonen (1990). "Socioeconomic Conditions in Childhood and Ischaemic Heart Disease During Middle Age," *BMJ*, Vol. 301, pp. 1121–23.

———— et al. (1988). "Social Connections and Mortality from all Causes and Cardiovascular Disease: Prospective Evidence from Eastern Finland," *Am J Epidemiol*, Vol. 128, pp. 370–80.

———— and J. E. Keil (1993). "Socioeconomic Factors and Cardiovascular Disease: A Review of the Literature," *Circulation*, Vol. 88, Part 1.

———— et al. (1996). "Inequality in Income and Mortality in the United States Analysis of Mortality and Potential Pathways," *BMJ*, Vol. 312, p. 999.

Kaplan, R.M. and E. J. Groessl (2002). "Applications of Cost-Effectiveness Methodologies in Behavioral Medicine," *Journal of Consulting and Clinical Psychology*, Vol. 70, pp. 482–93.

Karasek, R., T. Theorell, and J. Schwartz (1988). "Job Characteristics in Relation to the Prevalence of Myocardial Infarction in the U.S. Health Examination Survey and the Health and Nutrition Survey," *Am J Pub Health*, Vol. 78, pp. 1–9.

Kawachi, I. (1997). "Social Capital, Income Inequality, and Mortality," *American Journal of Public Health*, Vol. 87, pp. 1491–98.

———— et al. (1996). "A Prospective Study of Anger and Coronary Heart Disease: The Normative Aging Study," *Circulation*, Vol. 94, pp. 2090–95.

————, B. P. Kennedy, and R. Glass (1999). "Social Capital and Self-Rated Health: A Contextual Analysis," *Am J Pub Health*, Vol. 89, pp. 1187–93.

Keil, J.E. et al. (1984). "Incidence of Coronary Heart Disease in Blacks in Charleston, South Carolina," *Am Heart J*, Vol. 108 (Pt. 2), pp. 779–86.

———— et al. (1992). "Does Equal Socioeconomic Status in Black and White Men Mean Equal Risk of Mortality?," *Am J Public Health*, Vol. 82, pp. 1133–36.

Kelly, J.E. et al. (1997). "Health Effects of Emotional Disclosure in Rheumatoid Arthritis Patients," *Health Psychology*, Vol. 16, pp. 331–40.

Kendler, K.S. and C. O. Gardner (1997). "Religion, Psychopathology, and Substance Use and Abuse: A Multimeasure, Genetic-Epidemiologic Study," *Am J Psychiatry*, March, Vol. 154(3), pp. 322–29.

Kennedy, B.P., I. Kawachi, and D. Porthrow-Stith (1996). "Income Distribution and Mortality: Cross-Sectional Ecological Study of the Robin Hood Index in the United States," *BMJ*, Vol. 312, pp. 1004–7. See erratum: *BMJ*, Vol. 312, p. 1253.

Kiecolt-Glaser, J.K. et al. (1991). "Spousal Caregivers of Dementia Victims: Longitudinal Changes in Immunity and Health," *Psychosomatic Medicine*, Vol. 53, pp. 345–62.

——— et al. (1997). "Marital Conflict in Older Adults: Endocrinological and Immunological Correlates," *Psychosomatic Medicine*, Vol. 59, pp. 339–49.

Kilpatrick, D.G., P. A. Resick, and L. J. Vernonen (1981). "Effects of Rape Experience: A Longitudinal Study," *Journal of Social Issues*, Vol. 37, pp. 105–22.

King, A.C. et al. (1990). "The Relationship Between Repressive and Defensive Coping Styles and Blood Pressure Responses in Healthy, Middle-Aged Men and Women," *Journal of Psychosomatic Research*, Vol. 34, pp. 461–71.

Kington, R. and H. Nickens (2001). "Racial and Ethnic Differences in Health: Recent Trends, Current Patterns, Future Directions." In N. J. Smelser, W. J. Wilson, and F. Mitchell, eds. *America Becoming: Racial Trends and Their Consequences*, Vol. II, Washington, D.C.: National Academy Press, pp. 253–310.

Kitagawa, E.M. and P. M. Hauser (1973). *Differential Mortality in the United States: A Study in Socioeconomic Epidemiology.* Cambridge, Mass.: Harvard University Press.

Kleinke, C.L. and G. Williams (1994). "Effects of Interviewer Status, Touch, and Gender on Cardiovascular Reactivity," *Journal of Social Psychology*, Vol. 134, pp. 274–249.

Knowler, W.C. et al. (2002). "Reduction in the Incidence of Type 2 Diabetes with Lifestyle Intervention or Metformin," *N Engl J Med*, Vol. 346, pp. 393–403.

Koenig, H.G., (1999). *The Healing Power of Faith.* New York: Simon & Schuster.

———, D. O. Moberg, and J. N. Kvale (1988). "Religious Activities and Attitudes of Older Adults in a Geriatric Assessment Clinic," *Journal of the American Geriatric Society*, Vol. 36, p. 366.

———, J. N. Kvale, and C. Ferrel (1988). "Religion and Well-being in Later Life," *The Gerontologist*, Vol. 28, pp. 18–28.

——— et al. (1992). "Religious Coping and Depression Among Elderly, Hospitalized Medically Ill Men," *Am J Psychiatry*, Vol. 149, p. 1693.

——— et al. (1997). "Attendance at Religious Services, Interleukin-6, and Other Biological Parameters of Immune Function in Older Adults," *Int J Psychiatry Med*, Vol. 27 (3), pp 233–50.

————, L. K. George and B. L. Peterson (1998). "Religiosity and Remission of Depression in Medically Ill Older Patients," *Am J Psychiatry*, Vol. 155, pp. 536–42.

———— et al. (1998). "The Relationship Between Religious Activities and Blood Pressure in Older Adults," *Int J Psychiatry Med*, Vol. 28(2), pp. 189–213.

———— et al. (1998). "The Relationship Between Religious Activities and Cigarette Smoking in Older Adults," *J Gerontol A Biol Sci Med Sci*, Vol. 53(6), pp. M426–34.

———— and D. B. Larson (1998). "Use of Hospital Services, Religious Attendance, and Religious Affiliation," *South Med J*, Vol. 91(10), pp. 925–32.

Koskenvuo, M. et al. (1988). "Hostility as a Risk Factor for Mortality and Ischemic Heart Disease in Men," *Psychosom Med*, Vol. 50, pp. 153–64.

Krause, N. (1998). "Stressors in Highly Valued Roles, Religious Coping, and Mortality," *Psychology and Aging*, Vol. 13, pp. 242–55.

———— and Thanh Van Tran (1989). "Stress and Religious Involvement Among Older Blacks," *Journal of Gerontology: Social Sciences*, Vol. 44, pp. 4–13.

Kryzewski, M. and D. T. Phillips (2001). *Leading with the Heart: Coach K's Successful Strategies for Basketball, Business, and Life.* New York: Warner Books.

Kübler,-Ross E. (1969). *On Death and Dying.* New York: Macmillan.

———— (1987). *AIDS: The Ultimate Challenge.* New York: Collier Books.

Kubzansky, L.D. et al. "Breathing Easy: A Prospective Study of Optimism and Pulmonary Function in the Normative Aging Study," *Annals of Behavioral Medicine*, on press.

———— and I. Kawachi (2000). "Going to the Heart of the Matter: Do Negative Emotions Cause Coronary Heart Disease?," *Journal of Psychosomatic Research*, Vol. 48, pp. 323–37.

———— and I. Kawachi (1999). "Socioeconomic Status, Hostility, and Risk Factor Clustering in the Normative Aging Study: Any Help from the Concept of Allostatic Load?," *Ann Behav Med*, Vol. 21, pp. 330–38.

———— et al. (1998). "Is Educational Attainment Associated with Shared Determinants of Health in the Elderly? Findings from the MacArthur Studies of Successful Aging," *Psychosomatic Medicine*, Vol. 60, pp. 578–85.

———— et al. (2001). "Is the Glass Half Empty or Half Full? A Prospective Study of Optimism and Coronary Heart Disease in the Normative Aging Study," *Psychosomatic Medicine*, Vol. 63, pp. 910–16.

———— et al. (1998). "Anxiety and Coronary Heart Disease: A Synthesis of Epidemiological, Psychological, and Experimental Evidence," *Ann Behav Med*, Vol. 20, pp. 47–58.

Kuhn, C.M. and S. M. Schanberg (1998). "Responses to Maternal Separation: Mechanisms and Mediators," *Int J Dev Neurosci*, Vol. 16, pp. 261–70.

Landwirth, H. (1996). *The Gift of Life*. Give Kids World Foundation, Kissimmee, FL 34746, (407) 396-1114.

Lane, J.D. and D. M. Wegner (1995). "The Cognitive Consequences of Secrecy," *Journal of Personality and Social Psychology*, Vol. 69, pp. 237–53.

Langer, E. and R. Abelson (1974). "A Patient by Any Other Name . . . : Clinician Group Difference in Labeling Bias," *Journal of Consulting and Clinical Psychology*, Vol. 42, pp. 4–9.

———— and J. Rodin, (1976). "The Effects of Choice and Enhanced Personal Responsibility for the Aged: A Field Experiment in an Institutional Setting," *Journal of Personality and Social Psychology*, Vol. 34, pp. 191–98.

Lantz, P.M. et al. (1998). "Socioeconomic Factors, Health Behaviors, and Mortality," *JAMA*, Vol. 279, pp. 1703–8.

Larson, D.B. et al. (1988). "The Impact of Religion on Blood Pressure Status in Men," *Journal of Religion and Health*, Vol. 28, pp. 265–78.

———— et al. (1997). "Attendance at Religious Services, Interleukin-6, and Other Biological Parameters of Immune Function of Older Adults," *International Journal of Psychiatry in Medicine*, Vol. 27, pp. 233–50.

———— et al. (1998). "The Relationship Between Religious Activities and Blood Pressure in Older Adults," *International Journal of Psychiatry in Medicine*, Vol. 28, pp. 189–213.

Larson, J.S. (1996). "The World Health Organization's Definition of Health: Social Versus Spiritual Health," *Social Indicators Research*, Vol. 38, pp. 181–92.

Lazarus, R.S. (1994). *Emotion and Adaptation*. Cary, N.C.: Oxford University Press.

———— and S. Folkman(1984). *Stress, Appraisal, and Coping*. New York: Springer Pub. Co.

———— and B. N. Lazarus (1994). *Passion & Reason, Making Sense of Our Emotions*. Oxford: Oxford University Press.

LeDoux, J. (1996). *The Emotional Brain: The Mysterious Underpinnings of Emotional Life*. New York: Simon & Schuster.

Lehman, C.D. et al. (1991). "Impact of Environmental Stress on the Expression of Insulin-Dependent Diabetes Mellitus," *Behav Neurosci*, Vol. 105, pp. 241–45.

Lepore, S.J. (1997). "Expressive Writing Moderates the Relation Between Intrusive Thoughts and Depressive Symptoms," *Journal of Personality and Social Psychology*, Vol. 73, pp. 1030–37.

———— (1998). "Problems and Prospects for the Social Support-Reactivity Hypothesis," *Society of Behavioral Medicine*, Vol. 20, pp. 257–69.

————. "A Social-Cognitive Processing Model of Emotional Adjustment to Cancer." In A. Baum and A. Anderson, eds., *Psychological Interventions for Cancer.* Washington, D.C.: American Psychological Association, on press.

Lerner, M.J. (1980). *The Belief in a Just World.* New York: Plenum.

Leserman, J. et al. (1999). "Progression to AIDS: The Effects of Stress, Depressive Symptoms, and Social Support," *Psychosom Med,* Vol. 61, pp. 397–406.

———— et al. (2000). "Impact of Stressful Life Events, Depression, Social Support, Coping, and Cortisol on Progression to AIDS," *Am J Psychiatry,* Vol. 157, pp. 1221–28.

Levin, J.S., L. M. Chatters, and R. J. Taylor (1995). "Religious Effects on Health Status and Life Satisfaction Among Black Americans," *Journal of Gerontology: Social Sciences,* No. 3, pp. 158–61.

————, K. S. Markides, and L. A. Ray (1996). "Religious Attendance and Psychological Well-being in Mexican Americans: A Panel Analysis of Three Generations' Data," *The Gerontologist,* Vol. 36, pp. 454–63.

Levine, B. "A Harsh, Swift Clarity," *Los Angeles Times,* July 31, 2001, pp. E2–E4.

Lewinsohn, P.M. et al. (1980). "Social Competence and Depression: The Role of Illusory Self-Perception," *Journal of Abnormal Psychology,* Vol. 89, pp. 203–12.

Lewontin, R.C. (1973). "The Apportionment of Human Diversity," *Evolutionary Biology,* Vol. 6, pp. 381–98.

Lichtenstein, P. et al. (2000). "Environmental and Heritable Factors in the Causation of Cancer: Analyses of Cohorts of Twins from Sweden, Denmark, and Finland." *NEJM,* Vol. 343, pp. 78–85.

Lincoln, C.E., and L.H. Mamiya (1990). *The Black Church in the African American Experience.* Durham, N.C.: Duke University Press.

Litt, M.D. (1995). "Preparation for Oral Surgery: Evaluating Elements of Coping," *Journal of Behavioral Medicine,* Vol. 18, p. 435.

———— et al. (1992). "Coping and Cognitive Factors in Adaptation to In Vitro Fertilization Failure," *Journal of Behavioral Medicine,* Vol. 15, pp. 171–87.

————, C. Nye, and D. Shafer (1993). "Coping with Oral Surgery by Self-Efficacy Enhancement and Perceptions of Control," *J Dent Res,* Vol. 72, pp. 1237–243.

Lorig, K. and H. Holman (1993). "Arthritis Self-Management Studies: A Twelve-Year Review," *Health Education Quarterly,* Vol. 20(1), pp. 17–28.

———— et al. (1998). "Arthritis Self-Management Program Variations: Three Studies," *Arthritis Care and Research,* Vol. 11, pp. 448–54.

Lovallo, W. (1997). *Stress & Health: Biological and Psychological Interactions.* Thousand Oaks, Calif.: Sage Publications, Inc.

Low-Beer et al. (2000)."Health-Related Quality of Life Among Persons with HIV After the Use of Protease Inhibitors," *Quality of Life Research,* Vol. 9, pp. 941–49.

Lutgendorf, S.K. et al. (1994). "Changes in Cognitive Coping Strategies Predict EBV-Antibody Titre Change Following a Stressor Disclosure Induction," *Journal of Psychosomatic Research,* Vol. 38, pp. 63–78.

Lyman, R., "A Director's Journey into a Darkness of the Heart," *The New York Times,* June 24, 2001, pp. 23–24.

Lynch, J.W., G. A. Kaplan, and J. T. Salonen (1997). "Why Do Poor People Behave Poorly? Variation in Adult Health Behaviors and Psychosocial Characteristics by Stages of the Socioeconomic Lifecourse," *Soc Sci Med,* Vol. 44, pp. 809–19.

McCullough, M.E. et al. (2000). "Religious Involvement and Mortality: A Meta-Analytic Review," *Health Psychology,* Vol. 19, pp. 211–22.

McEwen, B.S. (1998). "Protective and Damaging Effects of Stress Mediators," *N Engl J Med,* Vol. 338, pp. 171–79.

——— and E. Stellar (1993). "Stress and the Individual: Mechanisms Leading to Disease," *Arch Int Med,* Vol. 153, pp. 2093–101.

McGee, D.L. et al. (1999). "Self-Reported Health Status and Mortality in a Multiethnic US Cohort," *Am J Epidemiol,* pp. 41–6.

McIntosh, D.N. (1995). "Religion-as-Schema, with Implications for the Relation Between Religion and Coping," *The International Journal for the Psychology of Religion,* Vol. 5, pp. 1–16.

———, R. C. Silver, and C. B. Wortman (1993). "Religion's Role in Adjustment to a Negative Life Event: Coping with the Loss of a Child," *Journal of Personality & Social Psychology,* Vol. 65, pp. 812–21.

McQuaid, E.L. and J. H. Nassau (1999). "Empirically Supported Treatments of Disease-Related Symptoms in Pediatric Psychology: Asthma, Diabetes, and Cancer," *Journal of Pediatric Psychology,* Vol. 24, pp. 305–28.

Malarkey, W.B. et al. (1994). "Hostile Behavior During Marital Conflict Alters Pituitary and Adrenal Hormones," *Psychosomatic Medicine,* Vol. 56, pp. 41–51.

Manning, M.M. and T. L. Wright (1983). "Self-Efficacy Expectancies, Outcome Expectancies, and the Persistence of Pain Control in Childbirth," *Journal of Personality and Social Psychology,* Vol. 45, pp. 421–31.

Manuck, S.B. et al. (1995). "The Pathogenicity of Behavior and Its Neuroendocrine Mediation: An Example from Coronary Artery Disease," *Psychosomatic Medicine,* Vol. 57, pp. 275–83.

Marmot, M. and M. J. Shipley (1996). "Do Socioeconomic Differences in Mortality Persist After Retirement? 25-Year Follow-up of Civil Servants from the First Whitehall Study," *BMJ*, Vol. 313, pp. 1177–80.

———— and G. Rose (1984). "Inequalities in Death: Specific Explanations of a General Pattern?" *Lancet*, Vol. 1, pp. 1003–6.

————, M. Kogevinas, and M. A. Elston (1987). "Social/Economic Status and Disease," *Ann Rev Public Health*, Vol. 8, pp. 111–35.

———— and G. Davey Smith (1997). "Socio-Economic Differentials in Health: The Contribution of the Whitehall Studies," *Journal of Health Psychology*, Vol. 2, pp. 283–90.

———— et al. (1987). "Employment Grade and Coronary Heart Disease in British Civil Servants," *J Epidemiol Comm Hlth*, Vol. 32, pp. 244–49.

———— et al. (1991). "Health Inequalities Among British Civil Servants: The Whitehall II Study," *Lancet*, Vol. 337, pp. 1387–93.

———— et al. (1997). "Contribution of Job Control and Other Risk Factors to Social Variations in Coronary Heart Disease," *Lancet*, Vol. 350, pp. 235–40.

Marmot, M.G.H., M. Bobak, and G. Davey Smith (1995). "Explanations for Social Inequalities in Health." In B. C. Amaick et al., eds. *Society and Health*. London: Oxford University Press.

Marshall, G.N. et al. (1992). "Distinguishing Optimism for Pessimism: Relations to Fundamental Dimensions of Mood and Personality," *Journal of Personality and Social Psychology*, Vol. 62, pp. 1067–74.

Martikainen, P. and T. Valkonen (1996)."Mortality After Death of Spouse in Relation to Duration of Bereavement in Finland," *Journal of Epidemiology and Community Health*, Vol. 50, pp. 264–68.

Maruta, T. et al. (2000). "Optimists vs. Pessimists: Survival Rate Among Medical Patients Over a 30-Year Period," *Mayo Clin Proc*, pp. 140–43.

Masten, A.S. (2001). "Ordinary Magic, Resilience Processes in Development," *American Psychologist*, Vol. 56, pp. 227–38.

Matarazzo, J. (1984). "Behavioral Health: A 1990 Challenge for the Health Services Professions." In Matarazzo, J. et al. *Behavioral Health: A Handbook of Health Enhancement and Disease Prevention*. New York: Wiley and Sons.

Maton, K.I. (1989). "The Stress-Buffering Role of Spiritual Support: Cross-Sectional and Prospective Investigation," *Journal of the Scientific Study of Religion*, Vol. 28, pp. 303–10.

———— et al. (1992). "Religious Coping and Depression Among Elderly, Hospitalized Medically Ill Men," *American Journal of Psychiatry*, Vol. 149, pp. 1693–1700.

———, L. K. George, and B. L. Peterson (1998). "Religiosity and Remission from Depression in Medically Ill Older Patients," *American Journal of Psychiatry*, Vol. 155, pp. 536–42.

Matthews, K.A. et al. (1989). "Educational Attainment and Behavioral and Biologic Risk Factors for Coronary Heart Disease in Middle-Aged Women," *American Journal of Epidemiology*, Vol. 129, pp. 1132–44.

Meaney, M.J. et al. (2000). "Postnatal Handling Increases the Expression of CAMP-Inducible Transcription Factors in the Rat Hippocampus: The Effects of Thyroid Hormones and Serotonin," *Journal of Neuroscience*, Vol. 20, pp. 3926–35.

Miller, J.J., K. Fletcher, and J. Kabat-Zinn (1995). "Three-Year Follow-up and Clinical Implications of a Mindfulness Meditation-Based Stress Reduction Intervention in the Treatment of Anxiety Disorders," *General Hospital Psychiatry*, Vol. 17, pp. 192–200.

Miller, T.W. et al. (1996). "Meta-Analytic Review of Research on Hostility and Physical Health," *Psychological Bulletin*, Vol. 119, pp. 322–48.

Miringoff, M.L., M. Miringoff, and S. Opdycke (2001). *The Social Report: A Deeper Understanding of Prosperity: Fordham Institute for Innovation in Social Policy*. Tarrytown, N.Y.: Fordham Graduate Center.

Musselman, D.L., D. L. Evans, and C. B. Nemeroff (1998). "The Relationship of Depression to Cardiovascular Disease," *Arch Gen Psychiatry*, Vol. 55, p. 580.

Myers, L.B. and C. R. Brewin (1994). "Recall of Early Experience and the Repressive Coping Style," *Journal of Abnormal Psychology*, Vol. 103, pp. 282–92.

———, R. Brewin, and M. J. Power (1998). "Repressive Coping and the Directed Forgetting of Emotional Material," *Journal of Abnormal Psychology*, Vol. 107, pp. 141–48.

National Human Genome Research Institute, *Five Year Strategic Plan for Reducing Health Disparities*, Area of Emphasis Number 1A, www.genome.gov.

National Institutes of Health (9/2001). "Toward Higher Levels of Analysis: Progress and Promise in Research on Social and Cultural Dimensions of Health: Executive Summary." Washington, D.C.: Office of Behavioral or Social Sciences Research, NIH Publication No. 21-5020. www.obsst.od. gov/ publications.

———. "Addressing Health Disparities: The NIH Program of Action." www. healthdisparities.hi.gov.

Nelson, G. (1989). "Life Strains, Coping and Emotional Well-Being: A Longitudinal Study of Recently Separated and Married Women," *American Journal of Community Psychology*, Vol. 17, pp. 459–83.

———— (1994). "Emotional Well-being of Separated and Married Women: Long-term Follow-up Study," *American Journal of Orthopsychiatry,* Vol. 64, pp. 150–60.

Niaura, R. et al. (1992). "Repressive Coping and Blood Lipids in Men and Women," *Psychosomatic Medicine,* Vol. 54, pp. 698–706.

Nolen-Hoeksema, S., J. S. Girgus, and M.E.P. Seligman (1992). "Predictors and Consequences of Childhood Depressive Symptoms: A 5-Year Longitudinal Study," *Journal of Abnormal Psychology,* Vol. 101, pp. 405–22.

————, J. Morrow, and B. L. Fredrickson (1993). "Response Styles and the Duration of Episodes of Depressed Mood," *Journal of Abnormal Psychology,* Vol. 102, pp. 20–28.

————, S.N., A. McBride, and J. Larson (1997). "Rumination and Psychological Distress Among Bereaved Partners," *Journal of Personality and Social Psychology,* Vol. 72, pp. 855–62.

Norem, J.K. and N. Cantor (1986a). "Anticipatory and Post Hoc Cushioning Strategies: Optimism and Defensive Pessimism in 'Risky' Situations," *Cognitive Therapy and Research,* Vol. 10, pp. 347–62.

———— (1986b). "Defensive Pessimism: Harnessing Anxiety as Motivation," *Journal of Personality and Social Psychology,* Vol. 51, pp. 1208–217.

Olds, D.L. et al. (1997). "Long-term Effects of Home Visitation on Maternal Life Course and Child Abuse and Neglect: Fifteen-Year Follow-up of a Randomized Trial, *JAMA,* Vol. 278, pp. 637–43.

O'Leary, A. (1985). "Self-Efficacy and Health," *Behav Res Ther,* Vol. 23, pp. 437–51.

———— et al. (1988). "A Cognitive-Behavioral Treatment for Rheumatoid Arthritis," *Health Psychology,* Vol. 7, pp. 527–44.

Oliver, M.L. and T. Shapiro (2001). "Wealth and Racial Stratification." In N. Smelser, W. Wilson, and F. Mitchell, eds., *America Becoming: Racial Trends and Their Consequence.* Washington, D.C.: National Academy Press.

Olshansky, S.J., B. A. Carnes, and C. Cassel (1990). "In Search of Methuselah: Estimating the Upper Limits to Human Longevity," *Science,* Vol. 250, pp. 634–40.

Oman, D., Ph.D., and D. Reed, M.D., Ph.D. (1998). "Religion and Mortality Among the Community-Dwelling Elderly," *American Journal of Public Health,* Vol. 88, p. 1469.

Ornish, D. (1999). *Love & Survival: 8 Pathways to Intimacy and Health.* New York: Harper-Collins.

Orth-Gomér, K. and J. Johnson (1987). "Social Network Interaction and Mortality: A Six-Year Follow-up of a Random Sample of the Swedish Population," *J Chronic Dis,* Vol. 40, pp. 949–57.

———— et al. (1998). "Social Relations and Extent and Severity of Coronary Artery Disease," *European Heart Journal*, Vol. 19, pp. 1648–56.

Ory, M. et al. (1999). "Prevalence and Impact of Caregiving: A Detailed Comparison Between Dementia and Nondementia Caregivers," *Gerontologist*, Vol. 39, pp. 177–85.

Ostrove, J.M. et al. (2001). "Objective and Subjective Assessments of Socioeconomic Status and Their Relationship to Self-Rated Health in an Ethnically Diverse Sample of Pregnant Women," *Health Psychology*, Vol. 19, pp. 613–18.

Otten, M. et al. (1990). "The Effect of Known Risk Factors on the Excess Mortality of Black Adults in the United States," *JAMA*, Vol. 263, pp. 845–50.

Pappas, G. et al. (1993). "The Increasing Disparity in Mortality Between Socioeconomic Groups in the U.S., 1960 and 1986," *New Eng Jrnl Med*, Vol. 329, pp. 126–27.

Pargament, K.I. (1997). *The Psychology of Religion and Coping: Theory, Research, Practice*. New York: The Guilford Press.

Park, C.L., L. H. Cohen, and R. L. Murch (1996). "Assessment and Prediction of Stress-Related Growth," *Journal of Personality*, Vol. 64, pp. 171–105.

———— and Susan Folkman (1997). "Meaning in the Context of Stress of Coping," *Review of General Psychology*, Vol. 1, pp. 115–44.

Parkes, C.M. (1975). "What Becomes of Redundant World Models? A Contribution to the Study of Adaptation to Change," *British Journal of Medical Psychology*, Vol. 48, pp. 131–37.

Pennebaker, J.W. (1989). "Confession, Inhibition, and Disease." In L. Berkowitz, ed., *Advances in Experimental Social Psychology*, Vol. 22. New York: Academic Press, pp. 211–44.

———— (1990). *Opening Up: The Healing Power of Confiding in Others*. New York: Morrow.

———— (1993). "Putting Stress into Words: Health, Linguistic, and Therapeutic Implications," *Behav Res Ther*, Vol. 31, pp. 539–48.

———— (1997). "Writing About Emotional Experiences as a Therapeutic Process," *Psychological Science*, Vol. 8, pp. 162–65.

———— and M. Francis (1996). "Cognitive, Emotional, Language Processes in Disclosure," *Cognition and Emotion*, Vol. 10, pp. 601–26.

———— and R. C. O'Heeron (1984). "Confiding in Others and Illness Rate Among Spouses of Suicide and Accidental-Death Victims," *Journal of Abnormal Psychology*, Vol. 93, pp. 473–76.

———— and S. K. Beall (1986). "Confronting a Traumatic Event: Toward an Understanding of Inhibition and Disease," *Journal of Abnormal Psychology*, Vol. 95, pp. 274–81.

———— et al. (1987). "The Psychophysiology of Confession: Linking Inhibitory and Psychosomatic Processes," *Journal of Personality and Social Psychology,* Vol. 52, pp. 781–93.

————, J. Kiecolt-Glaser, and R. Glaser (1988). "Disclosure of Traumas and Immune Function: Health Implications for Psychotherapy," *Journal of Consulting and Clinical Psychology,* Vol. 56, pp. 239–45.

————, S. D. Barger, and J. Tiebout (1989). "Disclosure of Traumas and Health Among Holocaust Survivors," *Psychosomatic Medicine,* Vol. 51, pp. 577–89.

————, T. J. Mayne, and M. E. Francis (1997). "Linguistic Predictors of Adaptive Bereavement," *Journal of Personality and Social Psychology,* Vol. 72, pp. 863–71.

———— and A. Graybeal (2001). "Patterns of Natural Language Use: Disclosure, Personality, and Social Integration," *Current Directions in Psychological Science,* Vol. 10, pp. 90–93.

Peterson, C. and M.E.P. Seligman (1984). "Causal Explanations as a Risk Factor for Depression: Theory and Evidence," *Psychological Review,* Vol. 91, pp. 347–74.

————, B. A. Bettes, and M.E.P. Seligman (1985). "Depressive Symptoms and Unprompted Causal Attributions: Content Analysis," *Behav Res Ther,* Vol. 23, pp. 379–82.

————, M.E.P. Seligman, and G. E. Vaillant (1988). "Pessimistic Explanatory Style Is a Risk Factor for Physical Illness: A Thirty-Five-Year Longitudinal Study," *Journal of Personality and Social Psychology,* Vol. 55, pp. 23–27.

———— et al. (1998). "Catastrophizing and Untimely Death," *Psychological Science,* Vol. 9, p. 127.

Petrie, K. J. et al. (1995). "Disclosure of Trauma and Immune Response to Hepatitis B Vaccination Program," *Journal of Consulting and Clinical Psychology,* Vol. 63, pp. 787–92.

Pickering, T. et al. (1988). "How Common Is White Coat Hypertension?" *Journal of the American Medical Association,* vol. 259, pp. 225–28.

Pincus, T., L. F. Callahan, and R. V. Burkhauser (1987). "Most Chronic Diseases Are Reported More Frequently by Individuals with Fewer Than 12 Years of Formal Education in the Age 18–64 United States Population," *J Chron Dis,* Vol. 40, No. 9, pp. 865–74.

Pollner, M. (1989). "Divine Relations, Social Relations, and Well-being," *Journal of Health and Social Psychology,* Vol. 30, pp. 92–104.

Power, C., O. Manor, and J. Fox (1991). *Health and Class: The Early Years.* London: Chapman & Hall.

Price, R. (2000). *A Whole New Life.* New York: Scribner.

Putnam, R.D. (2001). *Bowling Alone: The Collapse and Revival of American Community.* New York: Touchstone Books.

Raeikkoenen, K. et al. (1999). "Effects of Optimism, Pessimism, and Trait Anxiety on Ambulatory Blood Pressure and Mood During Everyday Life," *Journal of Personality and Social Psychology,* Vol. 76, pp. 104–13.

Ramachandran, V.S. (1994). *Encyclopedia of Human Behavior (Four-Volume Set).* Academic Press.

Redelmeier, D.A. and S. M. Singh (2001). "Survival in Academy Award-Winning Actors and Actresses," *Annals of Internal Medicine,* vol. 134, pp. 955–62.

Reed, G.M. et al. (1994). "Realistic Acceptance as a Predictor of Survival Time in Gay Men with AIDS," *Health Psychology,* Vol. 13, pp. 299–307.

———— (1999). "Negative HIV-Specific Expectations and AIDS-Related Bereavement as Predictors of Symptom Onset in Asymptomatic HIV-Positive Gay Men," *Health Psychology,* Vol. 18, pp. 354–63.

Reich, R. (2000). *The Future of Success.* New York: Alfred A. Knopf.

Rime, B. (1995). "Mental Rumination, Social Sharing, and the Recovery from Emotional Exposure." In J. W. Pennebaker, ed., *Emotion, Disclosure, and Health.* Washington, D.C.: American Psychological Association.

Rodin, J., and E. J. Langer (1977). "Long-term Effects of a Control-Relevant Intervention with the Institutionalized Aged," *Journal of Personality and Social Psychology,* Vol. 35, pp. 897–902.

Rogot, E. et al. (1992). *A Mortality Study of 1.3 Million Persons by Demographic, Social and Economic Factors: 1979–1985 Follow-up.* National Institutes of Health. NIH Publication No. 92–3297, pp. 1–5.

Rose, G. and M. G. Marmot (1981). "Social Class and Coronary Heart Disease," *BMJ,* Vol. 45, pp. 13–19.

Rosengren, A. et al. (1993). "Stressful Life Events, Social Support, and Mortality in Men Born in (1933)." *BMJ,* Vol. 307, pp. 102–5.

Rothbart, M., M. Evans, and S. Fulero (1979). "Recall for Confirming Events: Memory Processes and the Maintenance of Social Stereotyping," *Journal of Experimental Social Psychology,* Vol. 15, pp. 342–55.

Rowe, J.W. and R. L. Kahn (1998). *Successful Aging.* New York: Pantheon Books.

Rozanski, A., J. A. Blumenthal, and J. Kaplan (1999). "Impact of Psychological Factors on the Pathogenesis of Cardiovascular Disease and Implications for Therapy," *Circulation,* Vol. 99, pp. 2192–2217.

Ruberman W. et al. (1984). "Psychosocial Influences on Mortality After Myocardial Infarction," *N Engl J Med,* Vol. 311, pp. 552–9.

Ruehlman, L.S., S. G. West, and R. J. Pasahow (1985). "Depression and Evaluative Schemata," *Journal of Personality*, Vol. 53, pp. 46–92.

Russek, L.G. and G. E. Schwartz (1997a). "Feelings of Parental Caring Predict Health Status in Midlife: A 35-Year Follow-up of the Harvard Mastery of Stress Study," *Journal of Behavioral Medicine*, Vol. 20, pp. 1–13.

——— (1997b). "Perceptions of Parental Caring Predict Health Status in Midlife: A 35-Year Follow-up of the Harvard Mastery of Stress Study," *Psychosomatic Medicine*, Vol. 59, pp. 144–49.

Sanderson, W.C., R. M. Rapee, and D. H. Barlow (1989). "The Influence of an Illusion of Control on Panic Attacks Induced via Inhalation of 5.5% Carbon Dioxide-Enriched Air," *Arch Gen Psychiatry*, Vol. 46, pp. 157–62.

Sapolsky, R. (1994, 1998). *Why Zebras Don't Get Ulcers: An Updated Guide to Stress, Stress-Related Diseases, and Coping.* New York: W. H. Freeman and Company.

Saylor, C.F., C. C. Swenson, and P. Powell (1992). "Hurricane Hugo Blows Down the Broccoli: Preschoolers' Post-Disaster Play and Adjustment," *Child Psychiatry and Human Development*, Vol. 22, pp. 139–49.

Schaefer, J.A. and R. H. Moos (1992). "Life Crises and Personal Growth." In B. Carpenter, ed., *Personal Coping: Theory, Research, and Application.* Westport, Conn.: Praeger, pp. 149–70.

Schechter, M.T. et al. (1994). "Higher Socioeconomic Status Is Associated with Slower Progression of HIV Infection Independent of Access to Health Care," *J Clin Epidemiol*, Vol. 47, pp. 59–67.

Scheier, M.F., J. K. Weintraub, and C. S. Carver (1986). "Coping with Stress: Divergent Strategies of Optimists and Pessimists," *Journal of Personality and Social Psychology*, Vol. 51, pp. 1257–264.

——— et al. (1989). "Dispositional Optimism and Recovery from Coronary Artery Bypass Surgery: The Beneficial Effects on Physical and Psychological Well-being," *Journal of Personality and Social Psychology*, Vol. 57, pp. 1024–40.

——— (1999). "Optimism and Rehospitalization After Coronary Artery Bypass Graft Surgery," *Arch Intern Med*, Vol. 159, p. 829.

Scheier, M.J., C. S. Carver, and M. W. Bridges (1994). "Distinguishing Optimism from Neuroticism (and Trait Anxiety, Self-Mastery, and Self-Esteem): A Reevaluation of the Life Orientation Test," *Journal of Personality and Social Psychology*, Vol. 67, pp. 1063–78.

Schnall, L. and D. Baker (1994). "Job Strain and Cardiovascular Disease," *Ann Review Public Health*, Vol. 15, pp. 381–411.

Schnall, P., C. Pieper, and J. Schwartz (1990). "The Relationship Between 'Job Strain,' Workplace Diastolic Blood Pressure, and Left Ventricular Mass Index," *JAMA*, Vol. 263, pp. 1929–35.

Schoutrop, M.J.A. et al. (1996). "Overcoming Traumatic Events by Means of Writing Assignments." In A. Vingerhoets, F. van Bussel, and J. Boelhouwer, eds., *The (Non) Expression of Emotions in Health and Disease*. Tilburg, The Netherlands: Tilburg Press, pp. 279–90.

Schulz, R. and S. R. Beach (1999). "Caregiving as a Risk Factor for Mortality: The Caregiver Health Effects Study," *JAMA*, Vol. 282, pp. 2215–19.

——— et al. (1997). "Health Effects of Caregiving: The Caregiver Health Effects Study: An Ancillary Study of the Cardiovascular Health Study," *Annals of Behavioral Medicine*, Vol. 19, pp. 110–16.

——— et al. (1996). "Pessimism, Age, and Cancer Mortality," *Psychology and Aging*, Vol. 11, pp. 304–9.

Schwartz, C.E. et al. (1997). "The Quality-of-Life Effects of Interferon Beta-1b in Multiple Sclerosis: An Extended Q-Twist Analysis," *Arch Neurol*, pp. 1475–480.

Schwartzberg, S. (1994). "Vitality and Growth in HIV-Infected Gay Men," *Soc Sci Med*, Vol. 38, pp. 593–602.

Seeman, M. and S. Lewis (1995). "Powerlessness, Health and Mortality: A Longitudinal Study of Older Men and Mature Women," *Soc Sci Med*, Vol. 41, pp. 517–25.

Seeman, T. (1996). "Self-Efficacy Beliefs and Change in Cognitive Performance: MacArthur Studies on Successful Aging," *Psychology & Aging*, Vol. 11, pp. 538–51.

Seeman, T.E. et al. (1987). "Social Network Ties and Mortality Among the Elderly in the Alameda County Study," *American Journal of Epidemiology*, Vol. 126, p. 714.

——— (1994). "Social Ties and Support and Neuroendocrine Function: The MacArthur Studies of Successful Aging." *Annals of Behavioral Medicine*, Vol. 16, pp. 95–106.

———, M. L. Bruce, and G. J. McAvay (1996). "Social Network Characteristics and Onset of ADL Disability: MacArthur Studies of Successful Aging," *Journals of Gerontology: Series B. Psychological Sciences & Social Sciences*, Vol. 51B, pp. S191–S200.

Segerstrom, S.C. et al. (1996). "Causal Attributions Predict Rate of Immune Decline in HIV-Seropositive Gay Men," *Health Psychology*, Vol. 15, pp. 485–93.

——— et al. (1998). "Optimism Is Associated with Mood, Coping, and Immune Change in Response to Stress," *Journal of Personality and Social Psychology*, Vol. 74, pp. 1646–655.

Seligman, M.E.P. (1991). *Learned Optimism: How to Change Your Mind & Your Life*. New York: Pocket Books.

——— et al. (1988). "Explanatory Style Change During Cognitive Therapy for Unipolar Depression," *Journal of Abnormal Psychology*, Vol. 97, pp. 13–18.

———— et al. (1995). *The Optimistic Child*. New York: Houghton Mifflin.

Siegel, K. and E. W. Schrimshaw (2000). "Perceiving Benefits in Adversity: Stress-Related Growth in Women Living with HIV/AIDS," *Social Science and Medicine*, Vol. 51, pp. 1543–54.

Silver, R.C. et al. (2002). "Nationwide Longitudinal Study of Psychological Responses to September 11," *JAMA*, Vol. 288, pp. 1235–44.

Singer, B. and C. D. Ryff (1999). "Hierarchies of Life Histories and Associated Health Risks: Socioeconomic Status and Health in Industrial Nations." In N. Adler et al., eds., *Social, Psychological and Biological Pathways*. New York: The New York Academy of Sciences.

Smarr, K.L. et al. (1997). "The Importance of Enhancing Self-Efficacy in Rheumatoid Arthritis," *American College of Rheumatology*, Vol. 10, pp. 18–26.

Smith, J. (1999). "Healthy Bodies and Thick Wallets: The Dual Relation Between Health and Socioeconomic Status," *Journal of Economic Perspectives*, Vol. X, pp. 145–66.

Smith, T.W. (1992). "Hostility and Health: Current Status of a Psychosomatic Hypothesis," *Health Psychology*, Vol. 11, pp. 139–50.

————, P. C. Kendall, and F. J. Keefe (2002). "Behavioral Medicine and Clinical Health Psychology: Introduction to the Special Issue, a View from the Decade of Behavior," *Journal of Consulting and Clinical Psychology*, Vol. 70, pp. 459–62.

Smyth, J., N. True, and J. Souto (2001). "Effects of Writing About Traumatic Experiences: The Necessity for Narrative Structuring," *Journal of Social and Clinical Psychology*, Vol. 20, pp. 161–72.

Smyth, J.M. et al. (1999). "Effects of Writing About Stressful Experiences on Symptom Reduction in Patients with Asthma or Rheumatoid Arthritis: A Randomized Trial," *Journal of the American Medical Association*, Vol. 281, pp. 1304–9.

Snyder, C.R. and B. L. Dinoff (1999). "Coping: Where Have You Been?" in C. R. Snyder, ed., *Coping: The Psychology of What Works*. Oxford: Oxford University Press, pp. 3–19.

Spera, S.P., E. D. Buhrfeind, and J. W. Pennebaker (1994). "Expressive Writing and Coping with Job Loss," *Academy of Management Journal*, Vol. 37, pp. 722–33.

Spiegel, D., J. R. Bloom, and I. Yalom (1981). "Group Support for Patients with Metastatic Cancer: A Randomized Prospective Outcome Study," *Archives of General Psychiatry*, Vol. 38, pp. 527–33.

Stairfield, E. (1982). "Child Health and Social Status," *Pediatrics*, Vol. 69, pp. 550–57.

Stanton, A.L. and P. R. Snider (1993). "Coping with Breast Cancer Diagnosis: A Prospective Study," *Health Psychology,* Vol. 12, pp. 16–23.

Steffen, P.R. et al. (2001). "Religious Coping, Ethnicity, and Ambulatory Blood Pressure," *Psychosomatic Medicine,* Vol. 63, pp. 523–30.

Sternberg, E.M. (2000). *The Balance Within: The Science Connecting Health and Emotions.* New York: W. H. Freeman and Company.

Story, M., D. Neumark, and S. French (2002). "Individual and Environmental Influences on Adolescent Eating Behaviors," *Journal American Diet Asso,* Vol. 102, pp. S40–51.

Strawbridge, W.J. et al. (1997). "Frequent Attendance at Religious Services and Mortality over 28 Years," *American Journal of Public Health,* Vol. 87, p. 957.

Strogatz, D.S. et al. (1997). "Social Support, Stress, and Blood Pressure in Black Adults," *Epidemiology,* Vol. 8, pp. 482–87.

Strutton, D. and J. Lumpkin (1992). "Relationship Between Optimism and Coping Strategies in the Work Environment," *Psychology Reports,* Vol. 71, pp. 1179–186.

Surwit, R.S. et al. (1984). "Behavioral Manipulation of the Diabetic Phenotype in OB/OB Mice," *Diabetes,* Vol. 33, pp. 616–18.

Swann, W.B., Jr., and S. J. Read (1981). "Acquiring Self-Knowledge: The Search for Feedback That Fits," *Journal of Personality and Social Psychology,* Vol. 41, pp. 1119–28.

Taylor, S.E. (1983). "Adjustment to Threatening Events: A Theory of Cognitive Adaption," *American Psychologist,* Vol. 38, pp. 1161–73.

—— (1989). *Positive Illusions: Creative Self-Deception and the Healthy Mind.* New York: Basic Books.

—— (1994). "Positive Illusions and Well-Being Revisited: Separating Fact from Fiction," *Psychological Bulletin,* Vol. 116, pp. 21–27.

—— and J. D. Brown (1988). "Illusion and Well-Being: A Social Psychological Perspective on Mental Health," *Psychological Bulletin,* Vol. 103, pp. 193–210.

—— et al. (1992). "Optimism, Coping, Psychological Distress, and High-Risk Sexual Behavior Among Men at Risk for Acquired Immunodeficiency Syndrome (AIDS)," *Journal of Personality and Social Psychology,* Vol. 63, pp. 460–73.

—— et al. (2000). "Psychological Resources, Positive Illusions, and Health," *American Psychologist,* Vol. 55, pp. 99–109.

Tedeschi, R.G. et al. (1998). *Postraumatic Growth: Positive Changes in the Aftermath of Crisis.* Lawrence Erlbaum Associates, Inc.

Temoshok, L. (1987). "Personality, Coping Style, Emotion and Cancer: Towards an Integrative Model," *Cancer Surveys,* Vol. 6, pp. 836–37.

Theorell, T. et al. (1995). "Social Support and the Development of Immune Function in Human Immunodeficiency Virus Infection," *Psychosomatic Medicine,* Vol. 57, pp. 32–36.

Thompson, S.C. (1991). "The Search for Meaning Following a Stroke," *Basic and Applied Social Psychology,* Vol. 12, pp. 81–96.

Tix, A.P. and P. A. Frazier (1998). "The Use of Religious Coping During Stressful Life Events: Main Effects, Moderation, and Mediation," *Journal of Consulting and Clinical Psychology,* Vol. 66, pp. 411–22.

Uchino, B.N., J. T. Cacioppo, and J. K. Kiecolt-Glaser (1996). "The Relationship Between Social Support and Physiological Processes: A Review with Emphasis on Underlying Mechanisms and Implications for Health," *Psychological Bulletin,* Vol. 119, pp. 488–531.

———, D. Uno, and J. Holt-Lunstad (1999)."Social Support, Physiological Processes, and Health," *Psychological Science,* Vol. 8, pp. 145–48.

U.S. Department of Health and Human Services (1993). "Advance Report of Final Mortality Statistics," *Monthly Vital Statistics Report,* 42:5.

Vitaliano, P.P. et al. (1991). "Predictors of Burden in Spouse Caregivers of Individuals with Alzheimer's Disease," *Psychology and Aging,* Vol. 6, pp. 392–402.

Wadsworth, M.E.J. (1991). *The Imprint of Time: Childhood, History and Adult Life.* Oxford: Clarendon Press.

——— (1997). "Changing Social Factors and Their Long Term Implications for Health." In M. G. Marmot and M.E.J. Wadsworth, *Fetal and Early Childhood Environment: Long-term Health Implications.* London: Royal Society of Medicine Press Ltd.

Wagner, A.K. et al. (1995). "Advances in Methods for Assessing the Impact of Epilepsy and Antiepileptic Drug Therapy on Patients' Health-Related Quality of Life," *Quality of Life Research,* Vol. 4, p. 115.

Waite, L.J. and M. Gallagher (2001). *The Case for Marriage: Why Married People Are Happier, Healthier, and Better Off Financially.* New York: Broadway Books.

Waldron, I., M. E. Hughes, and T. L. Brooks (1996). "Marriage Protection and Marriage Selection—Prospective Evidence for Reciprocal Effects of Marital Status and Health," *Soc Sci Med,* Vol. 43, pp. 113–23.

Webster's New World Dictionary and Thesaurus (1998). New York: Simon & Schuster.

Wegner, D.M. (1997). "When the Antidote Is the Poison: Ironic Mental Control Processes," *Psychological Science,* Vol. 8, pp. 148–50.

——— and R. Erber (1992). "The Hyperaccessibility of Suppressed Thoughts," *Journal of Personality and Social Psychology,* Vol. 63, pp. 903–12.

———— and S. Zanakos (1993). "Ironic Processes in the Mental Control of Mood and Mood-Related Thought," *Journal of Personality and Social Psychology,* Vol. 65, pp. 1093–1104.

————, A. Broome, and S. J. Blumberg (1997). "Ironic Effects of Trying to Relax Under Stress," *Behav Res Ther,* Vol. 35, pp. 11–21.

———— and S. Zanakos (1994). "Chronic Thought Suppression," *Journal of Personality,* Vol. 62, p. 4.

Weinberger, D.A., G. E. Schwartz, and R. J. Davidson (1979). "Low-Anxious, High-Anxious, and Repressive Coping Styles: Psychometric Patterns and Behavioral and Physiological Responses to Stress," *Journal of Abnormal Psychology,* Vol. 88, pp. 369–80.

Welch-McCaffrey, D. et al. (1989). "Surviving Adult Cancers. Part 2: Psychosocial Implications," *Annals of Internal Medicine,* Vol. 111, pp. 517–24.

Wiedenfeld, S.A. et al. (1990). "Impact of Perceived Self-Efficacy in Coping with Stressors on Components of the Immune System," *Journal of Personality and Social Psychology,* Vol. 59, pp. 1082–94.

Wilcox, G.B. (1991). "Cigarette Brand Advertising and Consumption in the United States: 1949–1985," *Journal of Advertising Research,* Vol. 31, pp. 61–67.

Wilkenson, R.G. (1992). "Income Distribution Transition and Life Expectancy," *BMJ,* Vol. 304, pp. 165–68.

———— (1996). *Unhealthy Societies: The Afflictions of Inequality.* London: Routledge.

Williams, D. (2001). "Racial Variations in Adult Health Status: Patterns, Paradoxes, and Prospects." In N. Smelser, W. Wilson, and F. Mitchell, eds., *America Becoming: Racial Trends and Their Consequences.* Washington, D.C.: National Academy Press.

Williams, R.B. et al. (1992). "Prognostic Importance of Social and Economic Resources Among Medically Treated Patients with Angiographically Documented Coronary Artery Disease," *JAMA,* Vol. 267, pp. 267–520, 520–24.

Wilson, T.W. et al. (1993). "The Association Between Plasma Fibrinogen Concentration and Five Socioeconomic Indices in the Kuopio Ischemic Heart Disease Risk Factor Study," *Am J Epidemiol,* Vol. 137, pp. 292–300.

Wilson, W. (1987). *The Truly Disadvantaged: The Inner City, the Underclass, and Public Policy.* Chicago: University of Chicago Press.

Wing, R.R. et al. (2001). "Behavioral Science Research in Diabetes: Lifestyle Changes Related to Obesity, Eating Behavior, and Physical Activity," *Diabetes Care,* Vol. 24, pp. 117–23.

Winkleby, M.A., S. P. Fortmann, and D. G. Barrett (1990). "Social Class Disparities in Risk Factors for Disease: Eight-Year Prevalence Patterns by Level of Education," *Prev Med,* Vol. 19, pp. 1–12.

Woods, T.E. and D. B. Larson (1998). "Use of Hospital Services, Religious Attendance, and Religious Affiliation," *Southern Medical Journal*, Vol. 91, pp. 925–32.

——— et al. (1995). "Use of Acute Hospital Services and Mortality Among Religious and Non-Religious Copers with Medical Illness," *Journal of Religious Gerontology*, Vol. 9, pp. 1–22.

——— et al. (1999). "Religiosity Associated with Affective and Immune Status in Symptomatic HIV-Infected Gay Men," *Journal of Psychomatic Research*, Vol. 46, pp. 165–76.

Zullow, H.M. and M.E.P. Seligman (1990). "Pessimistic Rumination Predicts Defeat of Presidential Candidates, 1900 to 1984." *Psychological Inquiry*, Vol. 1, pp. 52–61.

——— et al. (1988). "Pessimistic Explanatory Style in the Historical Record: CAVing LBJ Presidential Candidates and East Versus West Berlin," *American Psycholog*, Vol. 43, pp. 673–82.

Index